Bikes and Bloomers

Bikes and Bloomers

Victorian Women Inventors and
Their *Extra*ordinary Cycle Wear

Kat Jungnickel

Goldsmiths
Press

© 2018 Goldsmiths Press
Published in 2018 by Goldsmiths Press
Goldsmiths, University of London, New Cross
London SE14 6NW

Printed and bound by Clays Ltd, St Ives plc
Distribution by The MIT Press
Cambridge, Massachusetts, and London, England

A CIP record for this book is available from the British Library

Library of Congress Cataloging-in-Publication Data
Names: Jungnickel, Katrina, author.
Title: Bikes and bloomers : Victorian women inventors and their
    extraordinary cycle wear / Kat Jungnickel.
Description: Cambridge, MA : Goldsmiths Press, [2018] |
    Includes bibliographical references and index.
Identifiers: LCCN 2017039950 | ISBN 9781906897758 (hardcover : alk. paper)
Subjects: LCSH: Women cyclists – Clothing – Great Britain – History – 19th century. |
    Women's clothing – Great Britain – History – 19th century. |
    Cycling – Social aspects – Great Britain – History – 19th century.
Classification: LCC GV1054 .J86 2018 | DDC 646.4/04–dc23
LC record available at https://lccn.loc.gov/2017039950

ISBN 978-1-906897-75-8 (hbk)
ISBN 978-1-906897-78-9 (ebk)

www.gold.ac.uk/goldsmiths-press

To Raylee, my mum, who taught me to sew,
tell stories and be brave in the world.

Most people who have followed the recent discussion of the dress question must have marveled at the rapid change in public opinion. It is only within the last few weeks that we have begun to realize that the time of our emancipation from a dangerous and incongruous riding skirt is at hand.

Of course, there will be a terrible outcry against it, and an attempt will be made to denounce it as immodest, but time will work wonders. Perhaps in time, even it will be understood that women should be judged by their lives and not by distinct dress.

All we can do now is persevere.

Ada Earland
Dress Reform For Women
Bicycling News, 1893

Contents

Acknowledgements

My writings about the past have been greatly inspired by people in the present. This book is for my friends in the vast, warm and wonderful cycling communities I have had the pleasure to be involved with over the past few years. They have supported this project in many ways; from attending events, asking questions and trying on costumes, to carting things around and setting up exhibitions. We have also gone on some great rides. My ideas are richer for their interest and encouragement. To LFGSS, London Bike Kitchen, LBK WAGFEST, Look Mum No Hands, CarryMe Cargo Bikes, The League of Ordinary Riders, YACF and more – thank you!

Like all projects of this nature, there are many people formative in its making over the last few years. I use 'we' a lot throughout the following pages, as the research would not have been possible without collaboration. This book is a product of the combined contributions of a collection of people. The core interdisciplinary team included: Rachel Pimm (research assistant) who was involved in almost every aspect of the project; Nadia Constantinou (pattern cutter) who translated our ideas and the patents into block patterns; and Alice Angus (artist) who worked in the studio/office with us, listening, drawing and painting in response to our findings. Alice's wonderful illustrations were digitally printed onto silk and sewn into skirts as linings, and are revealed when the garments convert from ordinary dress into extraordinary cycling wear. The linings add yet another layer to these remarkable multi-dimensional storytelling objects. Thanks also go to Guy Hill of Dashing Tweeds who supported the project and to Nikki Pugh, Annette-Carina van der Zaag and Britt Hatzius who helped with research, sewing and the exhibition, Charlotte Barnes who took many of the terrific photos and Lan-Lan Smith and Amanda Windle who appear in costume.

I also want to thank Michaela Benson and Imogen Tyler who reminded me at exactly the right time how social writing should be. To James Fraser, it almost goes without saying but also cannot be said enough, thank you for always being there. Many thanks to Genevieve Bell and Nina Wakeford who continue to encourage my practice and support my work, and to Rachel Aldred for the opportunity to start to explore an interest in Victorian cycle wear during my postdoc. My appreciation also goes to the Economic and Social Research Council and Intel Corporation for initial funding and to Goldsmiths Sociology Department, which is filled with an incredible group of intellectually generous people who support me in doing the kinds of critical and creative research that have inspired this book.

Finally, this project is dedicated to Raylee, my mum. She taught me to sew and showed me that if I could imagine it, I could make it. She instilled in me a love of creating three-dimensional objects sculpted out of material and mess, pins and patterns, time and sometimes blind optimism. After a break of many years, this project reawakened many embodied sewing memories. It was poignant, as it was a time when mum was slowly disappearing through illness. But it is through sewing that she stays with me. I am also greatly moved by and feel an awesome responsibility to the cycle wear inventors in this book. I hope I have done justice to their creative and inventive spirit and commitment to exploring new ways of being in the world.

Overall, this book is what happens when sewing, cycling and sociology collide. It is sometimes worrying to bring all your interests together in a project. As fond as I am of alliteration, I wondered if it might dent my passions. It didn't. In fact the opposite happened. I remembered my love for sewing, made new friends and networks and I now think differently with and through things. I also have a lot more cycle wear in the wardrobe. I wholeheartedly encourage my students to bring their interests to their work. After all, doing what you love can only make life better.

Part I

Introduction: Making, Wearing and Inventing Futures

This book is about Victorian engineering, feminist cultures of invention and new mobility technologies. More specifically, it is about women cycle wear inventors and their valuable contributions to cycling's past. There are abundant studies about the history of the bicycle and its technical trajectory. We seem to know a lot about what we have ridden over the years. We know less about what we have worn. And even less about the women who invented new forms of clothing specifically for cycling. The bicycle in late nineteenth-century Britain is often celebrated as a vehicle of women's liberation. This book focuses on another critical technology with which women forged new and mobile public lives – *cycle wear*.

I aim to make a contribution to the histories of British innovation and cycling with fascinating lesser-known tales about a group of inventive Victorian women who not only imagined, made and wore radical new forms of cycle wear – they also patented their cutting-edge designs. Inventors of this period used their differently clad bodies and the patent system to carve out new public identities. Some of them did this everyday on the streets while others travelled afar to cut international business deals. A few even cultivated celebrity status. Regardless of the nature or scale of their efforts, they all pushed at the parameters of established forms of mobile and gendered citizenship and forged new paths into social, political and economic worlds that together give shape to the urban landscape for today's generation of active women.

These stories make up a remarkable piece of history, made all the more so by how little known it is. The book is set in the 1890s, when a cycling craze swept the nation and many middle- and upper-class

Victorians enthusiastically took to the bicycle. Women quickly discovered that not only was society initially largely inhospitable to this new form of feminine mobility but their garments were vastly unsuitable. Fortunately for us, little was going to stop them cycling. The late nineteenth century also heralded a patenting boom, which together with the cycling craze set the scene for dramatic interventions in women's mobility clothing. While the activities of Dress Reformers, who campaigned for *rational* dress over *irrational* fashion to enable women to lead more active lives, are better known in this period, they were not the only Victorians responding to social and sartorial challenges and agitating for change. There were many others who sought to sew a way out of the dual 'dress problem' – which was how to cycle safely and comfortably yet also evade looking too much like a cyclist, so to minimise social hostility from onlookers who disproved of newly mobile women. Although motivations were wide-ranging, with some seeking business opportunities and others campaigning for women's rights, inventors were united by a collective desire to offer alternatives to restrictive notions of how newly mobile women could move in public.

My focus is on women, although men were also inventing women's cycle wear at this time, because solving the 'dress problem' was so mobilising that cycle wear inventions became a primary vehicle for women's entry into the world of patenting.[1] The sheer volume of their inventive capabilities in the mid 1890s rendered them statistically relevant for the first time (in Annual British Patent Reports), marking out new territory in what had previously been a predominantly masculine domain. Despite this, their stories are not well known. None of the patented garments in this book are on show in museums or galleries and few readers would be familiar with inventors' names. There are many reasons for this, including the fact that women's inventions have not garnered the historical recognition they deserve, as per gender norms of the time, and sewing has long been a feminine vocation, often underpaid and undertaken away from the public gaze. Also, as will become evident, the very nature of some of their secret sartorial cycling solutions were so effective in many cases that they were hidden during use and remained undetected long afterwards.

What can we learn from cycle wear? Quite a lot it turns out. Much like the bicycle, cycle wear mediates relationships between the body, other technologies, public space and society. It enables and also constrains mobility. Then, as now, over a century later, what an individual wears to move and the debate that surrounds it tells us a great deal about their rights as a citizen and place in society. In this case, Victorian cycle wear differed greatly for men and women, as did society's response to the sight of cycling women. And respond they did. The sheer volume and vitriolic nature of verbal and physical abuse directed at women on bicycles and dressed in cycle wear is illustrative of the threat many felt they posed to accepted conventions of the time. Studying cycle wear reveals how and in what ways different bodies fit (or in this case didn't fit) physically and also ideologically with changing ideas about being in and moving through public space. Although seemingly mundane, the study of cycle wear provides a rich and insightful lens into society-in-the-making.

Patents, similarly, are fascinating sources of social and technical data. Inventors tell us their names, where they lived, their vocations and, for women, married status. Each identifies a problem and solution, recommends materials and techniques, describes the target audience and suggests how their inventions could be put to use. Patents reveal how the politics of mobility and ideas around gender, citizenship and public space have been debated, imagined and materialised onto bodies over time. They give insights not only into social and material realities but also technological imaginaries – how inventors hoped to enact different futures. Usefully for us, patent archives record the work of big as well as small inventors, which means they do not have to be conventionally (as in commercially) successful to be valuable repositories of knowledge. Gaining first-hand access to people operating at the cutting-edge of social problems and solutions like this, and especially from women who in the past have had limited public voice, is rare and valuable. Reviewed critically, patents also raise questions about the nature and politics of historical knowledge and knowledge makers: who gets credited in the past and remembered in the present? Why do only some stories survive the test of time? What other factors shape our collective memories? And if that wasn't already enough, patents also provide handy detailed instructions for replicating inventions (more on this shortly).

While the patent archives hold much treasure, for me the most remarkable clothing patents of this time were for *convertible cycle wear*. These unique designs specifically respond to the 'dress problem' by enabling wearers to secretly switch ordinary clothing into cycle wear, and back again when required. Because the aim was to remain undetected, ingenious engineering is deliberately hidden in hems and seams. These costumes feature a range of concealed devices for getting material out of the way of wheels and pedals. Weighted pulleys, gathering straps, complex button-and-loop systems and more are built *into* the infrastructure of skirts. Convertible cycle wear represents a remarkable form of Victorian invention, much of which was designed by and for women, for the purpose of providing freedom of movement and, above all, new forms of power and control. Designs like this gave women *choice* over when, where and how they cycled. Inventors also firmly grasped, via patenting, the means to claim their ideas in the public domain. These stories add much needed layers and textures to cycling histories because they depict women as politically engaged citizens actively driving social and technical change.

Unable to locate existing material examples of these unique British inventions, we set about to (re)make a collection of them, taking inspiration from detailed step-by-step instructions in the patents. The process of making and wearing patented cycle wear transformed two-dimensional documents into three-dimensional arguments. Viewing patents as historical hackables enabled us to enter into dynamic time-travelling dialogues with inventors by literally getting into their inventions. In previous work I have termed this 'making things to make sense of things'.[2] Patents, archives and the office/studio became ethnographic fieldsites and sewing became a means of interviewing women who lived over a century ago. This book, as a result, presents a mixed-methods account of cycling, sewing and suffrage. It combines archival materials, patent studies and cycle wear designs with genealogical stories about inventors, their influences and lives pieced together with an ethnography of making and wearing a collection of convertible cycle wear.

In many ways this is a classic feminist reclamation project. Rendering women inventors' stories visible via archival materials and patented garments is a means of bringing them into multiple forms of being that cannot be easily overlooked. It is a deliberate and strategic countermove given that these costumes were designed not to be seen, were made by a marginalised group who were economically and legally under-represented and are largely absent from official records. It is about re-placing women inventors (back) into their designs, their patents into the chronicles of cycling and technology and (our) bodies into their costumes.

A quick note about what this book is not. Critically, this is not a historical re-enactment project. As I argue throughout, there was no *single* motivation driving these inventors, nor was there a definitive design solution to the challenges faced by Victorian cycling women, and similarly we did not set out to make a perfect replica of each patented garment. By making our own versions of each patent (five full garments comprising 27 items in total) we set out to explore and experience inventors' ideas and inventions in multi-dimensional forms. There are many ways to interpret, unpick and stitch these materials together and understand them in embodied practice. What I present in the following pages is but one account of the rich, complex, contradictory, highly textured and multi-faceted lives of inventive women who lived before us.

One thing is more certain – what women wear to cycle *still* matters. Contemporary cyclists may well find many of the concerns that plagued early women cyclists in the 1890s all too familiar – skirts that caught in pedals and wheels, inappropriate cuts and fabrics for active lifestyles, a limited range of cycles and cycle wear, media that ignored or trivialised women's achievements and, on top of that, a hostile public who catcalled and abused them as they cycled by. Given these experiences unfortunately continue to shape cycling for many today, inventors' solutions from over a century ago hold surprising relevance. Sewing cyclists might be happy to see that we have made patterns inspired by the patent research available for further creative interpretation.[3] Perhaps some of the inventors' insightful technical and material responses will once again find a home in contemporary cyclists' wardrobes.

Outline of the Book

Structurally, the book takes three parts. It starts by exploring women's freedom of movement in the Victorian era to identify what motivated many to engage so directly with cycling's 'dress problem'. I discuss changes to the British patenting system that enabled a broader group of inventors, including women, to claim their inventions and I review some of the strategies they used to forge new ways of being in and moving through public space with their differently clothed bodies. This section ends with an overview of the breadth of inventive outputs produced in the last decade of the nineteenth century.

The second section is concerned specifically with patents, biographies and materials. Tracing the lives of six inventors and their convertible costumes takes us on a journey through time and across Britain. Chapter 6 starts with *Alice Louisa Bygrave*, a Dressmaker from Brixton, south London, and her 1895 patent for 'Improvements in Ladies' Cycling Skirts' which features a dual pulley system sewn inside the front and rear seams of the skirt. Chapter 7 is about *Madame Julia Gill*, a Court Dressmaker of Haverstock Hill, north London, and her patent, also lodged in 1895, for 'A Cycling Costume for Ladies'. This ordinary A-line skirt gathers up to the waist into what she calls 'semi-skirt' via rings and ribbons hidden under a decorative flounce. *Frances Henrietta Müller*, a Gentlewoman from Maidenhead, Berkshire, and her 1896 patent for 'Improvements in Ladies' Garments for Cycling and other Purposes' is the focus of Chapter 8. She invented a three-piece convertible cycling suit made of a tailored coat, long skirt that folds to the waist via loops and buttons and a combined vest and knickerbocker undergarment. Chapter 9 is about sisters from Yorkshire, *Mary Elizabeth and Sarah Ann Pease*, whose 1896 patent for an 'Improved Skirt, available also as a Cape for Lady Cyclists' gave wearers the flexibility of removing the skirt altogether and wearing it as a cape when required. The final chapter is about *Mary Ann Ward* from Bristol who in 1897 patented 'Improvements in Ladies' Skirts for Cycling'. This skirt has rows of side buttons and waist straps that converted it into what was known as the 'Hyde Park Safety Skirt'.

Each chapter combines archival and genealogical materials and a key theme emerging in the research (such as collaboration, experimentation, activism, entrepreneurialism and in/formal networking) and ends with ethnographic insights garnered from sewing costumes. In the final section I reflect on why we know so little about women's inventive contributions to cycling's past and how we can do better.

Throughout the book, the intimacy of making and wearing the clothes of others shapes how I listen and give voice to my subjects. Readers will note that I use inventors' first names when writing about their lives, influences and designs. This is a deliberate political provocation and world-making claim. In a patriarchal society, women's last names are not their own. Historical documents render all too visible the dominance of men through time and even when women feature, they are blind to the lineage of matriarchal kinship. To continue this practice serves to reproduce this erasure. Instead the book follows in the footsteps of many, including nineteenth-century French seamstresses who, as Maria Tamboukou notes, 'signed their articles with their first names only, as a mode of rebellion against the name of the father or the husband that was imposed on them.'[4]

Finally, while this is a critical account that draws from multiple sources and is located within a feminist and science and technology studies theoretical framework, my writing is clearly imbued with respect and admiration for these women's creative responses and radical imaginings, and in many cases, sheer single-minded determination and unwavering courage. Learning about past lives invites us to reflect on our own. Perhaps we too can channel the energy that surfaces from social mistreatment into some small yet vivid intervention and play a role in making life better for future generations. What clearly emerges from all of these stories is how ideas make change. No invention is too small. No garment is too trivial. Clare Hemmings has written about how she 'always loved feminist theory for its utopianism' and 'dogged optimism that allows its practitioners to understand and experience life differently.'[5] I also write in this spirit, in the hope that these women's stories might become a catalyst to re-write our own.

1

'One Wants Nerves of Iron': Cycling in Victorian Britain

Minnie says 'Oxford is the most begotted place in the ~~world~~ kingdom and the meeting is likely to raise a great protest in the papers which will deter followers'. It certainly cannot be worse to ride in Oxford than in London, especially London suburbs. It's awful – one wants nerves of iron. The shouts and yells of the children deafen one, the women shriek with laughter or groan and hiss and all sorts of remarks are shouted at one, occasionally some not fit for publication. One needs to be very brave to stand all that. It makes one feel mad and ones ideas of humanity at large sink to a very low standard. When one gets out into the country there is little trouble beyond an occasional shout, but it takes some time to get away from these miles of suburban dwellings.

Letter from Kitty J. Buckman to Uriah, August 23, 1897[1]

Kitty loved cycling but given the fact that it took 'nerves of iron', it clearly wasn't always a relaxing or indeed safe pastime in late nineteenth-century Britain. This is an excerpt from one of many letters she and her cycling companions – Minnie, Jane, Uriah and Maude – wrote to each other about their experiences. In 1897 Kitty was 23 years old, and the younger sister of Sydney Savory Buckman, also known as S. S. Buckman, a renowned British geologist and supporter of the Rational Dress Movement. Kitty and her friends were keen cyclists and, fortunately for us over a hundred years later, ardent letter writers. Every missive describes recent or planned rides in and around Dorking, Hammersmith, Rickmansworth, Cirencester, Chippenham and Shrewton amongst other south-western English towns. We learn much about what these women loved about cycling in the 1890s. They tell lively stories of the weather, length and landscape of journeys, state of the roads, cost and quality of meals and accommodation, details of companions, mechanical issues, and increasing skills and fitness. Yet, as evoked in this aging cursive script, cycling was also often fraught with verbal and sometimes even physical assault. Why was this? Why did some women get treated in this way? What were Kitty and her friends doing and wearing that elicited such social violence? And how did they respond?

At the turn of the last century, new mobility technologies like the bicycle were seen as quintessential symbols of modernity and imaginings of the future. The shifts they promised for unchaperoned mobility and new social encounters were eagerly embraced; especially by women. The bicycle greatly expanded the lives of many, opening up 'a wide world to women, who have hitherto only seen places which they could reach in a carriage or by rail, or the places within walking distances of their homes'. It gave them 'a taste of independence, and much desired access broader social worlds and physical freedoms'.[2]

Yet, some bodies initially *fitted* better with this technology than others – some had to be made to fit. What women wore to lead more active lives and move freely in public space became one of many fronts in a landscape of hard-fought battles in which many sought to address discrimination and bias that limited their gender. This book is located in a small historical period spanning 1890 to 1900. This was a time of radical social, cultural and technological change and it sets a vivid and dynamic backdrop for the story of early cycle wear.

A cycling craze swept through Victorian society in the 1890s. Although the velocipede had been around in various iterations since the early nineteenth century, it was the 'Safety Bicycle' (in the form of diamond and step-through frames still in use today) that took the nation by storm. Prices were dropping, which made cycling accessible to a broader market, the invention of pneumatic tyres ensured a more pleasurable ride and machines were being actively marketed at women. While cycling was initially the preserve of middle- and upper-class citizens, a range of ages took to it and many evangelised its benefits. In 1895 Frances Willard of the Women's Christian Temperance Unit learnt to ride a bike she named 'Gladys' at the age of 53 and wrote *A Wheel Within a Wheel* to encourage others to do the same. 'I always felt a strong attraction toward the bicycle', she explains, 'because it is the vehicle of so much harmless pleasure'. Frances could not 'see a reason in the world why a woman should not ride the silent steed so swift and blithesome.'[3] However, even within this popular wave, it was still considerably easier for middle- and upper-class men to embrace this new modern means of moving. Socially accepted ideas about masculine mobility and clothing conventions meant that male bodies fitted much easier with bicycle technology. Women's dress, combined with the highly defined gendered norms that shaped their engagement with technology and public space were much more complicated.

Up until the mid 1800s, clothing was expensive and tailor-made, handed down or repurposed. It was a rare and precious commodity. According to Diana Crane in *Fashion and its Social Agenda*: 'Until the Industrial Revolution and the appearance of machine-made clothing, clothes were generally included among a person's most valuable possessions.'[4] To have a garment specially made for sport or dedicated physical activity was unthinkable for anyone apart from the aristocracy. It required not only disposable income, but also the leisure time to enjoy it. Most people's wardrobes were limited and shaped according to vocation, class, gender, religion and place of birth. As a result, what you wore in public revealed a great deal about your life and firmly located you in the social spectrum. However, the latter part of the century brought with it opportunities to make, buy and wear new kinds of clothing, and with it the chance to press against and re-define some of these conventional parameters.

The availability of sewing machines played a key role in this change. Although patented in 1851, the Singer Sewing Machine's popularity boomed in the 1890s, dominating the world market.[5] Compared to hand sewing, the machine vastly reduced the time it took to construct garments. Sewing patterns were becoming more widely distributed and targeted to specific consumers, such as Butterick's *New Styles for Bicycling*.[6] New machined materials such as wool, jersey and silk were becoming more widely accessible and were snapped up by cyclists for their beneficial properties in active wear. Added to this was a growing interest and awareness of the possibilities of other kinds of mobility cultures and clothing. News and images about social events at home and abroad were accelerating with the growth of mass media in the form of newspapers and popular periodicals. This further fuelled an appetite for new ideas and unusual goods that had been generated by the Great Exhibition of 1851 in Hyde Park, London, which was the first international showcase of manufactured products from around the world, and also by travellers who entertained eager listeners with tales from Grand Tours.

This was also a time of dress reform for both sexes. The Dress Reform Movement of the 1830s gained renewed drive in Britain in the 1890s in line with the growing popularity of sports and active lifestyles, particularly for women. Broadly, supporters sought to free Victorian women from the restrictions of what was termed 'ordinary' fashion, which largely took the form of tightly laced corsets, tailored blouses and jackets, floor-length skirts and layers of heavy petticoats. With a core aim of 'utility', rational dress comprised a range of styles but was ostensibly recognised as a looser corset (or no corset at all), a shortened skirt (or no skirt) and a bifurcated garment such as bloomers or knickerbockers. The Rational Dress Society, founded in 1881 and led by Lady Florence Harberton, was one such organisation that believed ordinary fashions were not only uncomfortable for everyday wear and any kind of physical activity, but also dangerous to health. Amongst other problems, they became waterlogged in rain, dragged in the dirt and caught on things. Newspapers regularly reported nasty incidents caused by errant skirts, such as women catching alight from discarded matches or being snagged by carriages and dragged along roads.[7] A writer in *The Rational Dress Gazette: Organ of the Rational Dress League* voiced concern with clothing's gendered disadvantage: 'Do we ever hear of a man being caught by his dress, or that his clothing catches fire? When will women awake to the dangers of skirts?'[8] On a more mundane level, conventionally clothed women often needed assistance to dress, enter and exit carriages, carry goods and care for children. As such, many believed that ordinary women's fashions, and skirts in particular, inhibited personal, social and political freedoms.

Like Kitty, many Victorian women enthusiastically took to the bicycle. Cycling, however, highlighted the problems of ordinary fashion more than any other sport or activity. Tightly fitted garments restricted breathing and excess material caught in pedals and wheels. Yet, it was not necessarily safer to dress in more cycle-oriented attire, as Kitty's letters attest. Kitty, her friends and brother were strong supporters of rational dress. Their letters provide insights for contemporary readers into what it might have been like to cycle in the 1890s in public in bloomers and without skirts. Although she persisted, and developed 'nerves of iron' along the way, Kitty empathised with others who might not be so dedicated. 'I don't wonder now in the least so many women having given up the R. D. [Rational Dress] Costume and returned to skirts.'[9]

The fact that cycling took place in public gained much needed exposure for the Dress Reform Movement but also greatly divided reactions. Some parts of society felt threatened by the sight of women cyclists carving out new controversial forms of gendered freedom in public space and responded with verbal and sometimes physical abuse. Even within the movement, there were some that wondered 'how far dress reform will be of benefit to cycling' and if cycling was a 'mere playful handmaiden to dress reform'.[10] Although emancipists had been agitating for women's rights throughout the century, the 1890s brought about a significant shift across a spectrum of social, political and economic spheres. Women were marching for the rights to education, work, equal pay and the vote. While many argued that women's bodies required release from constrictive costumes as much as they needed freedom from social restraints, not everyone felt the same. Responses were mixed given how much hostility women's cycling generated.

Despite this, many women bravely persevered. It was not long before women's cycle wear became visual shorthand for the 'New Woman'

Figure 1.1 Some of the popular Victorian periodicals that regularly covered happenings in the cycling world – *The Lady Cyclist*, *The Hub* and *Bicycling News*

A WEATHER FORECAST.

Mater Familias: *Skirts are all very well perhaps, when one's hands are free to hold them.*

Figure 1.2 An illustration of some of the practical problems of women's clothing – in the rain, carrying goods and looking after children, *The Rational Dress Gazette*, 1899

who was identified by her desire for progress, 'independent spirit and her athletic zeal'.[11] For some this was an attractive offering, and satirical magazines mocked women for buying and wearing the costume even if they did not own bicycles. Yet, this was not an easy identity to sustain in public. Women unchaperoned, moving at speed and in different places and times of the day than was familiar, shocked the establishment. This was not just about cycling. For some parts of society, these women were seen as rejecting their 'natural role', challenging the very essence of Victorian life and many of the taken-for-granted assumptions underpinning the home and masculine power. Therefore, to be seen cycling or in cycle wear of any kind in public required a great deal of courage to deal with responses ranging from gratuitous attention to verbal and physical abuse. To grasp what this might have been like, it helps to know the extent to which women cycling in ordinary dress suffered abuse. In a letter to the editor of the *Daily Mail*, two wheelwomen describe the struggle for *all* women cyclists, regardless of dress.

Figure 1.3 'Visiting Gown', an example of high-class fashion, *The Queen*, 1896

It is not only ladies who wear the rational costume who are hooted and abused in the London streets. My friend and I bicycle in ordinary everyday dress, yet we never go out without receiving a fair amount of this so-called chaff ... and occasionally we have had caps, etc, thrown at us.[12]

For those committed to cycle wear, there was also the burning issue of *what kind* of costume to wear and where to get it. Not only did women have to work up the nerve to convince those nearest to them of the value of the idea, they had to source an appropriate costume which was not easy. A cycling costume in the 1890s required a not insubstantial investment of money. Before the wider availability of ready-to-wear clothing,

Figure 1.4 A rationally dressed cyclist passes through town with 'Dignity' in the face of 'Impudence' from local men, *The Lady Cyclist*, 1896

it also involved much labour and time; women had to assemble ideas, designs or patterns, then make or adapt existing garments themselves or commission a sympathetic dressmaker or tailor – and not all were willing to support the idea of this newly mobile woman. To make such a garment could be interpreted as supporting a subversive political act. As Sarah Gordon writes in *'Any Desired Length': Negotiating Gender though Sports Clothing, 1870–1925*: 'With notions of gender so deeply embedded in clothing, changes in styles portended changes in the social structure.'[13]

How should it look? What should it do? There was no *one* answer. Everyone had an opinion on women's cycle wear. It was a hotly debated topic. Countless supporters and detractors made a heady mix. Even amongst dress reformers, the Rational Dress League, women's rights and associated emancipation groups and emerging cycling clubs, there was precious little consensus. So entrenched were the rules of what women should wear in Victorian society at specific times and places that the ensuing ambiguity provided minimal guidance for early adopters. Instead, pressure was placed on each cyclist to source and wear a garment that not only fitted and worked well on *and* off the bike but also embodied feminine charm in order to defuse all manner of social criticisms. It was not a simple undertaking. As Lady B warns in a regular *Lady Cyclist* column, 'Why Lady Cyclists Should Always Dress Well':

> The day of feminine cycling is yet young, wherefore it behoves all those in whose hands the moulding of popular prejudice and opinion is left to do their utmost to disarm the too ready criticisms of their self-appointed mentors.[14]

Men, Cycling and Cycle Wear

It is striking to consider tensions inherent in women's cycle wear in relation to the development of men's cycling and clothing. Middle- and upper-class men occupied a privileged position in Victorian society, dominating many aspects of public, political and commercial life. Bicycle riding presented little exception. It was initially considerably simpler for men to embrace this new modern means of moving. Social norms around masculine mobility combined with their clothing made for an easier fit. Victorian high-class men's fashion needed little adapting, compared to equivalent women's styles: it already encompassed narrow trousers, shirts and waistcoats, jackets, lace-up shoes and hats. In general men's clothing permitted more mobility than women's. Writing about the development of sportswear, Patricia Campbell Warner explains how men 'had simplified their dress for sports early on beginning in the mid-nineteenth century, when they softened their shirt collars, donned the new rubber-soled shoes, shucked off their jackets, wore knickerbockers or other loose trousers'.[15] Few of these freedoms were permitted to women of the same class. Although a new market for men's cycle wear swiftly emerged, it was nevertheless possible for a man in non-cycle-specific clothing to ride an Ordinary (commonly known as the Penny Farthing) or a Safety bicycle.

The Cyclists' Touring Club (CTC)[16] produced one of the first examples of dedicated cycle wear for men in 1882. The CTC uniform comprised a tailored jacket, breeches or knickerbockers, waistcoat, shirt, gaiters, helmet, polo caps, Deerstalker or Wideawake (soft felt hat), Puggarrees (thin scarf wrap) for helmets, cap covers, stockings and gloves. Members were allowed to simplify this to a 'jacket, breeches or knickerbockers, stockings and cap or helmet'.[17] Patterns were sent to a list of official tailors who provided the garments and designs were fiercely regulated: 'No braiding, epaulettes or trimmings under any circumstances be permitted upon the uniform', and any introductions of 'sweeping innovations' were dealt with swiftly. There was a responsibility not only to uphold the image of the club but also the respectability of cycling.

This uniform identified the male cyclist on and off the bicycle. The *Cyclists' Touring Club (CTC) Monthly Gazette* reported that the uniform for men 'continues to be a marvelous success'.[18] Many middle- and upper-class men were quick to adopt this visual symbol because identifying as a cyclist was in many circles a social achievement imbued with cultural cachet. However, men's cycle wear was not without anxieties at this time. Letters and columns in periodicals and newspapers point to issues concerning cost, what to do with an old uniform and discomfort in warm weather. It was also not without its critics. There was social consternation associated with this sartorial shift. A columnist in the upper-class women's magazine, *The Queen, The Lady's Newspaper*, acknowledges that '[e]very variation, however, slight, from the conventional costume always meets this treatment at the hands of the most ignorant of the British public'. But for men, this hostility was short-lived and much smaller in scale in comparison to the violent wrath directed at women cyclists. 'They were for a few weeks exposed to the same treatment that the lady cyclists have recently encountered', but now 'men can walk about in knickerbockers without attracting either offensive criticism or even eliciting a remark'.[19]

Partly, this was because cycling reflected and reproduced the social mores of a highly masculine society. Bicycles were initially expensive to buy and, in the case of the Ordinary bicycle, notoriously difficult to ride, especially on dirt roads and cobbles. For many, the unsafe qualities of high-wheel velocipedes were attractive features, and young, agile men were quick to assume this new exciting identity. Although cycling was a popular leisure activity, it was largely driven by the daring accomplishments of sporting heroes, who were almost always male. New models were often launched with epic endurance or speed record attempts. The 'Excelsior' Tricycle was advertised in the *CTC Gazette* as the 'Easiest Running, Lightest, Strongest and Fastest machine yet introduced' with 'PROOF! PROOF! PROOF! – More Races have been won upon the EXCELSIOR TRICYCLE than any other make'.[20] The bicycle was commonly discussed in terms of mastery, speed and distance, thereby reflecting and reinforcing Victorian ideals of industrial progress and modernity in association with athletic male bodies.

The popularity of cycle racing further refined men's cycle wear, and reinforced cycling's masculine appeal. In some cases there was very little material separating man from machine. Advertisements featured racing men in tight-fitting matching shorts and tops; a prelude to contemporary lycra skinsuits. Racing events took place in specially built velodromes that drew mass spectatorship. A slew of cycling periodicals, such as *The Hub*, *The Wheeler* and *Bicycling News*, emerged in response to the growing appeal of racing and celebrated cycling's masculine achievements. They published dramatic accounts of remarkable feats and showcased illustrations and photographs of bare-chested male winners in collectable commemorative posters. A discourse of heroism powerfully linked the male body and cycling with 'speed', 'agility', 'performance', 'champion', 'professional' and 'skill'. Cycle racing further demonstrated an easy coherence with and reinforcement of masculine forms of mobile citizenship. As Jennifer Hargreaves in *Sporting Females: Critical Issues in the History and Sociology of Women's Sports*, writes:

> Sports constituted a unique form of cultural life; they were overwhelming symbols of masculinity and chauvinism, embodying aggressive displays of physical power and competitiveness. In the nineteenth century there was no question that sports were the 'natural' domain of men and that to be good at them was to be essentially 'masculine'.[21]

Figure 1.5 Representations of men racing Ordinaries and Safety bicycles, *The Hub* and *The Wheeler*

Figure 1.6 Front page banner of *Cyclists' Touring Club Monthly Gazette*, 1883

So accepted was this 'natural' convergence between man and machine that even tragic accounts of male cycling deaths and crashes tended to reinforce their technical competence. Journalists often laid blame on technical malfunction or road surface over rider mistake. Rarely was men's skill or clothing at fault.

> An exceedingly sad and harrowing tricycle accident, whereby a Mr A.A. Broad met with his death, has just occurred at Croydon, not far from his own residence. It seems that the deceased gentleman was a riding a new, but faultily constructed, rear steering tricycle ... The deceased was comparatively a novice at the art, but even had he been an expert, the type of machine he rode would sooner or later have disastrously failed him – *CTC Gazette*, April 1884.[22]

> Mr. Lickfett, who has recently come to reside in Devonshire Road, went out for his maiden effort on a tricycle. He appeared to go on pretty well, but unwisely or unknowingly, he ventured his machine down the Tyson Road, where the gradient is 1 in 4 or 5. All control over the machine was gone in a moment as it rattled at a tremendous pace down the hill, precipitating the rider violently into the road [...] but for his robust constitutions, it is confidently believed, the accident would have terminated fatally – *CTC Gazette*, May 1884.[23]

The 'Dress Problem'

Unlike men, women were initially not so easily able to assume a safe or comfortable cycling identity through their clothing. Their ordinary fashions were not so straightforwardly fitted to the bicycle, there wasn't a single accepted uniform that they could adopt and they were not able to race like men and simply discard useless and non-functional attire. Women's fashion, and the social structures that shaped it, on the whole was considered more complex and difficult to adapt to the mobile challenges presented by the bicycle. The ongoing tensions inherent in the 'dress problem' were brought to life in discussions and debates about the cycling uniform in the *CTC Gazette*: 'Ladies in all parts of the kingdom have contributed their quote to the correspondence columns, and it is quite evident, from the half-concealed impatience of many of the fair sex, that they are burning to solve the vexed problem one way or another.'[24]

While the men's CTC uniform launched in the early 1880s and was shortly afterward declared a 'success', the women's version took many years, discussions, letters, exhibitions and meetings to resolve. Tailors, health professionals, clergy and cyclists put forth many differing opinions and designs and the CTC provided criteria by which they should be judged:

(1) Absolute freedom of movement for all parts of the body
(2) As great lightness as is compatible with warmth, with both lightness and warmth equally distributed. To which we might add 'a quiet and unobtrusive appearance'.[25]

The CTC held the opinion that women's cycle wear should not be too 'obtrusive'. By this it was possibly making reference to, and attempting to distance itself somewhat from more prominent suffrage activities, and in particular the public persona of Lady Harberton and the increasingly maligned Rational Dress Movement. The *Gazette* featured several updates over the following years indicating that it was no closer to solving the issue due to 'the likes and dislikes of the fair yet fickle sex'.[26] The lively debate was stabilised to a degree by 1884 when a design, featuring merino stockings and loose knickerbockers worn under a plain skirt with a tailored jacket and hat in Club cloth, was considered a 'popular' option for women cyclists.[27] However, it is important to note that it was designed for women cyclists riding high-wheel tricycles. It was not shaped around the complexity of sitting astride a more compact diamond frame or step-through two-wheeled bicycle, which was yet to come.

Although often referred to as the 'dress question' or the 'dress problem', the issue was clearly much larger than the costume. The difficulty in stabilising a design for women cyclists wasn't perhaps due so much to the fickleness of the wearers or the challenges of the bicycle or even environmental conditions but rather emblematic of far greater challenges that women faced in attempting to claim new vehicles of emancipatory practice. It was not simply a case of making minor adaptations to existing fashion but involved social and moral debates about the role of women in public space. Should women have independent mobility, mechanical competencies and attending social freedoms? What would this look like? What might it do to society?

Why Study Clothing?

Clothing is a mundane and essential feature of everyday life. We all need and wear clothes. It is central to how we protect ourselves from the elements and also how we perform, organise and make sense of society, each other and ourselves. Clothing is both a barometer and catalyst of social change. It is political, cultural and (still in many ways stubbornly) gendered. It is made and shaped in particular places and it also places people. You only have to wear the wrong thing to feel its exclusionary power and vice versa. For centuries, according to Diana Crane, it has also operated as a 'form of social control' with economic and symbolic value produced via specific identifiable work garments, as well as uniforms and other dress codes.[28] As a result, she argues that clothing can provide valuable insights into 'how people in different eras have perceived their positions in social structures and negotiated status boundaries.'[29] While middle- and upper-class men, for the most part, were able to more easily adapt their clothing to new forms of mobility, such as the bicycle, the case was not the same for women. Men did suffer social stigma, but it was not as violent or prolonged as that directed at women cyclists. Early women cyclists found that their conventional clothed bodies didn't fit so easily with new mobility technology, physically or socially. This motivated some, as we saw with Kitty and her friends, to use clothing as a means of carving out new ways for women to be in and move at speed in public. Yet, this was not without friction.

Despite its mundane ubiquity, or perhaps because of it, the study of clothing has not generated the attention it deserves. Patricia Campbell Warner counters the argument that the study of clothing is somehow too everyday or trivial to merit in-depth attention by reminding us that it is precisely because it 'can be a headache, a source of anxiety or self-consciousness, a cause of despair, a reason for envy, a focus for contention, a wrap of anonymity'[30] that we need to study it. Some also argue that clothing and fashion have been traditionally overlooked and understudied in western societies because they have been firmly embedded in the domain of women. In the absence of being able to own property or live independent lives, many women have creatively and imaginatively used clothing for purposes of self-expression and identity; power and practicality; loyalty and resistance. As a result, the study of clothing is particularly important for understanding women's lives. And, as I hope to show, inventive clothing practices by and for women are critical sites for making sense of the past and thinking about the present.

Of course, cycling and cycle wear were not the only means through which women were embracing new active lifestyles and challenging Victorian conventions. Many middle- and upper-class women had been engaging in a range of outside activities up until this point. Yet, there were stark differences between the expectations and demands of each. For instance, some were keen swimmers and gymnasts and, while these activities required customised clothing, their bodies were largely contained and concealed within bounded spaces; the water in the pool and the walls of the gymnasium. Women had also been horse riding for a long time, a pastime that shares more similarities with cycling in that it was located outside. Yet, and while there were exceptions, it mostly took place away from urban centres and intense social scrutiny. The custom to ride sidesaddle in long, layered habits also concealed a woman's legs and as such did not directly challenge the establishment in the way that cycling did. The mechanics of the bicycle required independently moving legs, heralding new and distinct design challenges to women's clothing and social norms. This, in part, helped to generate the conditions for invention. Victorians were keen inventors and the late nineteenth century was a time of immense creative experimentation across many disciplines; it brought forth many inventions from mundane artefacts such as toilet paper to devices of mass industrialisation, like electricity. It is well known that the bicycle proved to be an exciting catalyst for Victorian patenting. What is less known, is how the bicycle also motivated people, and particularly women, to invent and patent radical new forms of cycle wear.

The study of patents, as I discovered and hope to convey, is far from dry or boring. Patents get us closer to the experience of Victorian women cyclists because we can hear their voices, see parts of society through their eyes and trace their skillsets through the choreography of materials at play. This is unusual, because first-person historical accounts about women by women are rare. Their contributions, for many reasons, are often 'hidden from history'.[31] Patents therefore serve as a powerful tool to counter invisibility, because they provide an invaluable 'continuous source of information about market-related activities of women'.[32] In these historical artefacts inventors detail the problems they seek to fix and provide step-by-step instructions for future users to replicate their proposed solutions. We gain glimpses of where they were living and a sense of their daily lives. They tell us what they cared about and why. Their descriptions and images reveal social norms about how women's bodies were expected to move. We also see into potential futures, how they imagined themselves and others moving in and through public space in new ways. Here, for instance, is how Helena Wilson, a London Costumier, explains her new cycle wear invention in 1897:[33]

> This invention has for object a new or improved garter and dress distender for the use of ladies when cycling, riding on horseback and in similar outdoor exercises, where it is found expedient to dispense with the usual description of undergarments and to replace them by knickerbockers worn below a plain skirt.

Helena appears to be a supporter of dress reform for active women. However, she identifies a dual problem that is both socially and physically shaped:

> The latter dress although eminently adapted for the purposes described has the disadvantage of frequently and undesirably outlining the shape of the lower limb by too closely clinging to them, especially when travelling at some speed against a head-wind, and also impeding the free movement of the legs as required for pedalling and the like.

Her invention sought to address both issues, via what are effectively fluffy cuffs worn around the knee to obfuscate the shape of independently

Figure 1.7 Illustration accompanying Helena Wilson's patent

moving legs and simultaneously prevent the skirt from sliding, while not inhibiting the cyclist. In addition to explaining the state of dress reform and social concerns, patents provide information about available materials and manufacturing processes. The accompanying illustrations are similarly illuminating. Even though Helena's patent is only for a small item of dress, her illustrations place it in the context of a full cycling ensemble complete with a hat, jacket, skirt, gloves, bifurcated undergarment, stockings, shoes and spats. The unique cross-section cut-away reveals inside and outside the skirt. The drawing of the bicycle is also informative of associated technologies (such as step-through designs, mud guards and skirt guards) being added to velocipedes to accommodate women cyclists.

Most compelling of all, patents are critically valuable sites of data because they provide clear evidence of women actively driving change. They were not passively waiting for the situation to be resolved. They were not simply buffeted by social waves of change. Rather, they were actively attempting to drive it. By making and declaring their designs in public they became important actors in socio-technical change; legitimizing women's cycle wear as valid inventions and claiming a place for women in business.

While some might question if Helena's invention was particularly emancipatory for women, her design addressed criticisms directed at women's cyclists and in the process identified a potentially marketable product to solve it. Arguably, her aim was to keep women cycling, despite society's displeasure of catching an unsightly glimpse of their visibly moving legs. The fluffy cuffs were hidden under an ordinary skirt – no one would know they were there, if they did their job effectively. Alongside physically protesting in the streets or cycling in rational dress as Kitty and her friends were doing, the act of patenting cycle wear like this can be seen as a tactic for social and political change. It was an activity, which as we will see for different reasons, also required 'nerves of iron'.

2

From the Victorian Lady to the Lady Cyclist

'Fortunately,' he said, demonstrating the ideological underpinnings of many doctors' arguments, 'most women are engaged in house-keeping duties, and except for the want of open air, housework is probably the healthiest occupation a woman can have'.[1]

Patricia Vertinsky
The Eternally Wounded Woman: Women, Doctors,
and Exercise in the Late Nineteenth Century

To understand the many challenges facing early women cyclists, and what motivated them to materially respond, it is important to locate them in the framework of Victorian life and the many shifts taking place at the turn of the last century. Ideas around citizenship and freedom of movement are closely aligned with power, gender and public and private space. In Victorian society, women's mobility was physically and ideologically shaped by social codes and behavioural norms and closely linked to their clothing. In this chapter I trace the shift from Victorian Lady to Lady Cyclist.

Up until the mid-nineteenth century, the moral responsibility of reproduction, the bearing of and caring for children, defined middle- and upper-class women's lives. There were strict gender and labour divisions in place, whereby 'men produced, women reproduced'.[2] The lives of these women were deeply rooted in one particular place – the home. It was an inward-looking private and domestic role, distinctly different from that of the public life of men. According to Robinson's *The Art of Governing a Wife* (1747) women were 'to lay up and save; look to the house; talk to few; take of all within,' while men were encouraged 'to get; to go abroad and get his living; deal with other men; to manage all things without doors.'[3] Men were about being out in the world, while women's worlds were very much inside, which led to the accepted idea, as explained by Patricia Vertinksy in the quote at the beginning of this chapter, that housework provided more than ample exercise.

Movement for most middle- and upper-class women beyond the boundaries of the home was not only difficult in these socio-material confines but further inhibited by the pathologising of their health. It was a commonly accepted belief in early Victorian times that a woman had only a limited supply of energy and that she should conserve it to protect her reproductive potential. Many experts believed that exercise was unnecessary, and even potentially detrimental to their ability to reproduce, an opinion reinforced by the medical link between a woman's brain and her womb. The accepted notion of women as the 'weaker sex' came in part from the idea that they were disabled by their reproductive organs. Patricia Vertinksy has deeply researched the origins of this long-standing belief in her book *The Eternally Wounded Woman*. She writes: 'Since a woman's chief function was motherhood, the laws of nature demanded that not only should a bountiful energy supply be reserved for reproductive demands, but that more energy still should be earmarked to compensate for the monthly menstrual drain.'[4] Earmarked energy in this context was so broadly defined that it included the efforts of women who challenged social and gendered norms by reading too much, wanting an education, not wanting to marry or otherwise demanding more freedoms and forms of mobility than those allocated to her gender and class.

It was a difficult position to challenge. Becoming agitated as a result of feeling dissatisfied by this arrangement could result in a diagnosis of hysteria. This was a convenient catch-all label for troublesome women, and often came with terrible consequences. Elaine Showalter's *The Female Malady* tells harrowing tales of women whose desire for independence often found them labelled and punished. Her vivid account about 'women, madness and English culture from 1830' explores in detail the perceived threats to social order catalysed by women's desire to break free from repressive patriarchal society and the horrific ramifications of these actions: 'During an era when patriarchal culture felt itself to be under attack by its rebelling daughters, one obvious defense was to label women campaigning for access to the universities, the professionals, and the vote as mentally disturbed, and of all the nervous disorders of the *fin de siècle*, hysteria was the most strongly identified with the feminist movement.'[5]

Hysteria was a diagnosis that could catalyse a spectrum of traumatic treatments from the 'rest cure' to admission to an asylum. The former entailed being kept in bed and isolated from all stimulus until the woman calmed down and became more resolved to her life course. Charlotte Perkins Gilman, an American utopian feminist renowned for her novels, fiction and non-fiction, and social reform activities, tells us much about this treatment. In 1892 she wrote the now famous short story *The Yellow Wallpaper*, which presents an account of the rest cure being administered to a woman by her medical husband after she was diagnosed with an unnamed nervous disorder. Contemporary readers have recognised her symptoms as postpartum depression. Rather than providing support or care, or perhaps as a perverse form of it, her husband shuts her away from family, friends and the outside world. The story ends with the protagonist slowly descending into psychosis, as illustrated by her obsession with the room's yellow wallpaper. Although it is a fictional account, Charlotte was prescribed this treatment and the story apparently draws upon her personal experience. Charlotte's doctor, Silas Weir Mitchell, who also happened to be the rest cure's inventor, explains the extent of this immobilising treatment:

> In carrying out my general plan of treatment in extreme cases it is my habit to ask the patient to remain in bed from six weeks to two months. At first, and in some cases for four or five weeks, I do not permit the patient to sit up, or to sew or write or read, or to use the hands in any active way except to clean the teeth. In some instances I have not permitted the patient to turn over without aid ... In all cases of weakness, treated by rest, I insist on the patient being fed by the nurse.[6]

Although it was prescribed for many reasons, from anorexia nervosa to hysteria, the rest cure was administered to more women than men and could be used as a means of breaking the will of outspoken, emotional or independent individuals. Some have called the cure 'sadistic, controlling and intrusive'.[7] Virginia Woolf wrote about her experience of a version of the rest cure. She managed to avoid being bedridden but was sent to stay with an aunt in the country, at which she was less than happy. 'I have never spent such a wretched eight months in my life.'[8] Others were even less fortunate, such as Edith Lanchester, an honours student influential in a number of local political groups, who in 1895 was kidnapped and committed to a private asylum by her father and brothers on an 'urgency order'. The doctor in charge recorded the cause of her insanity as 'over-education'.[9] The extremity of these cures gives contemporary readers a sense of the dangers posed by women who did not conform to social and gendered norms. The fact that immensely influential women, such as those listed above, were prescribed these treatments points to the kinds of feminine behaviours frowned upon at the time. Also, that few were safe from its reach. As Elaine Showalter so clearly articulates, the 'hungry look' in the faces of rest cure patients 'was a craving for more than food'. These women 'were ravenous for a fuller life than their society offered them, famished for the freedom to act and to make real choices'.[10]

What catalysed such fear? Today we think about gender as something that is constructed and performed on a daily basis and not as something you are born with. Early Victorian life[11] was shaped by a more fixed understanding of gender as something that was biologically inherited and this widespread belief was often used to exclude women from political, cultural and intellectual life. An example of this is how the 'primacy of the reproductive organs in women was used to support the notion that a woman's major responsibility was to propagate the race'.[12] Therefore, if a person appeared to be abandoning their 'natural' gender, they were also abandoning their role in the family unit, and society more broadly, and for women this was a rejection of the very essence of pro-creation. 'The dominant point of view', writes Diana Crane in *Fashion and Its Social Agendas*, 'allowed for no ambiguity about sexual identification and no possibility for evolution or change in the prescribed behaviours and attitudes of each gender'.[13] Together, these labels and treatments and the rise of the Victorian madwoman in the late nineteenth century tell us much about social anxieties of the time and why the idea of new independent mobile women threatened to unravel the very fabric of society.

Clothing of the time produced and reinforced many accepted beliefs. It signalled respectability, decency and class: exactly where a person was socially located, what was expected of them and how they should be treated. For middle- and upper-class women, fashions comprised up to seven pounds of heavy skirts and layers of petticoats, binding corsets, tailored jackets, fitted blouses and veils. Together, these restrictions on the Victorian lady's body communicated clear class markers and also produced the perception, if not the reality in some cases, of a largely immobile citizen, considered incapable of much physical or indeed mental, economic or political movement. She was 'bound by the code of behavior as tight as the stays she was compelled to wear'.[14] Combined with apparent medical weaknesses, and legal and economic marginalisation, women faced a plethora of barriers to the public sphere, producing, as Elaine Showalter writes, 'one of history's self-fulfilling prophecies'. Social attitudes and treatments were 'used as a reason to keep women out of the professions, to deny them political rights, and to keep them under male control in the family and the state'.[15] It is no wonder that unsettling the gendered conventions of clothing sent shockwaves through deeply held beliefs. Clothing was not to be trivialised or under-estimated. Changes in clothing signalled changes in society.

Women's desire for new kinds of mobile freedom also usurped the foundations of respectable masculinity. In Victorian society, immobile women, or ladies of leisure, were viewed as the pinnacle of the feminine ideal and symbol of masculine achievement. To be free of work was indicative of higher class. It implied economic freedom. To *have* to work was considered degrading not only for the woman but for her husband as well. He was not accomplished enough if he failed to support his dependents. As Lee Holcombe writes in *Victorian Ladies at Work*, the term 'working ladies' had degrading connotations at a time when 'leisured, or idle, wives and daughters had become expensive status symbols for successful middle-class men'.[16] Clothes reinforced the idea that while women might work, ladies did not. Lavishly designed garments were a not-so-subtle indication that the wearer did not engage in manual labour. Layered restrictive clothing, with tightly laced corsets, also made it physically difficult, if not impossible, to do everyday domestic activities such as

cleaning or even walking any distance. The wearer's highly decorated immobility alluded to the wealth and status of the household –
she clearly had servants who undertook mundane tasks for their mis-
tress. Diana Crane argues that the 'restrictive conception of women's
roles' was 'reflected in the ornamental and impractical nature of fashionable clothing styles' which in turn 'contributed to the maintenance
of women in dependent, subservient roles.'[17]

There were of course notable exceptions at both ends of the class
spectrum. Lower-class women's ability to undertake hard physical
labour was indicative of the potential of robust female capacity. Yet,
their mobility in all other social and cultural contexts was differently
curtailed. While there are many examples of working women who
embraced practical garments related to their employment, as Diana
Crane notes, the public spaces where they 'violated Victorian norms for
clothing behavior ... were frequently "invisible" to the middle classes,
permitting these women to wear trousers and other items of masculine
clothing.'[18] There were also higher-class women and members of the
aristocracy who were able to shake off some of the stultifying restrictions
of home when they travelled abroad. Continental Grand Tours offered
a welcome relief to some from the confinements of English society.
Mary Berry, writing to her friend Gertie Greathead in 1798, proclaimed:
'Most thoroughly, do I begin to feel the want of that shake of our English
ways, English whims, and English prejudices, which nothing but leaving England gives one.'[19] However, even these women who enjoyed
more masculine freedoms of movement and had opportunities to create alternative gender narratives in the form of travel writing were still
encumbered by clothing restrictions, which were harder to shake. New
geographic and social freedoms were not always translated in clothing
'as women were forced to take "home" with them as they moved.'[20] The
Victorian Lady might well have escaped the confines of society at home
but she was unable to fully escape these feminine ideals abroad, as the
respectability they imbued provided the means through which she was
awarded these freedoms.

The 'New Woman' and Changing Social Fabric

By the late nineteenth century, however, things were changing. The suffrage movement was gaining traction, campaigning for women's freedom of movement out of the home and into education, business and politics. Some agitated for women's emancipation from patriarchal control on all fronts. Others took a less radical approach, claiming that educating women, and granting them admission to the workforce and healthy outdoor sports, actually prepared them better for marriage, to run complex households and produce healthy children. So, 'far from subverting homes, and family life, would actually improve them.'[21] Either way the aim was similar – to encourage society to accept independent mobile women. There was also change brewing in the medical profession. Doctors who had previously vigorously denounced women's physical activity started to cautiously encourage it. Irrespective of political platforms and leanings, opinion makers were becoming united on a shared desire for women to lead more active lives and clothing played a pivotal role in this.

Masses of middle- and upper-class Victorian women eagerly embraced new opportunities by taking up swimming, golf, fencing, gymnastics, tennis and athletics, as well as cycling. The sheer delight of being able to move freely was difficult to contain for some women. *The Hub* exclaimed: 'Many women seem to have gone daft over wheeling.'[22] Humorous accounts of women going missing from their domestic, wifely and daughterly duties started to appear in periodicals:

> A gentleman recently bought his wife and two daughters a bicycle apiece. It was not very long before he had occasion to regret his generosity. Returning home late one night he was annoyed to find the house deserted, Mary Jane out and no supper prepared.[23]

Active lifestyles demanded new active clothing and with it, as Sarah Gordon writes, an opportunity 'to produce a new conception of what it meant to be feminine.'[24] Women's clothing became a site for negotiation around what was possible, with many imagining and experimenting with new ways of being in and moving through public space. However, the fact that cycling was undertaken in urban streets and parks meant that it garnered much more attention than other popular sports. Located outside, it was uncontained and unbounded by sporting's rules and regulations. In contrast, swimming was encouraged, because here new radically (un)clothed female bodies could at least be partially concealed from the critical social gaze, in water within the bounds of the pool or by segregated seaside areas. Claire Parker writes about how 'the water always hid the effort of female swimmers and spectators rarely got close enough to see a swimmer in any discomfort' so there was no visible 'sweat, bulging muscles or signs of distress'. This tactical concealment led to swimming becoming accepted as a 'uniquely appropriate feminine sport'.[25] Interestingly, British women's style of swimwear was considered more relaxed than their American sisters, which, as Patricia Campbell Warner explains, had much to do with 'the English segregation of men and women at the seaside until the end of the century'.[26]

Although public debate around the importance of exercise for women was gathering momentum, medical professionals were quick to remind women of their weakness and limitations. Articles in popular cycling periodicals, commonly written by male doctors, propagated the notion that *too* much exercise would directly lead to injuries to women's health. Columns titled 'Should Women Cycle? A Medical View of the Question' warned women that they 'must remember that they belong to a sex which for centuries has not been accustomed to prolonged exercise in the open air and therefore must act with that discretion and caution which would often be quite unnecessary on a mans part'.[27] Others spoke of more ambiguous threats to a woman's health: 'Every woman who rides a wheel should understand that she can do so in moderation only, and if she attempts more she will pay for it dearly. The penalties may not be inflicted this year or next, but they are bound to come.'[28] The idea that women had a limited pool of energy that should not be wasted was an enduring trope used to regulate how and in what ways women could move in public space. Yet around these articles espousing moderation arose an enthusiasm for cycling and, with it, a dynamic new consumer culture.

The Construction of the Lady Cyclist

Women were fervent participants in the cycling craze that swept the nation in the late nineteenth century. The aristocracy started to cycle in the 1890s, which legitimised it as a respectable activity, and emboldened even timid cyclists to take to the streets and parks. The popularity of the sport was such that many businesses eagerly plied their wares to this new consuming public. Cyclists quickly became ideal test subjects for new products. They had already demonstrated a predilection for new cutting-edge technologies, had disposable income and their bodies were out in the elements. They also read and enthusiastically talked at length about all things related to cycling (which will come as no surprise to contemporary cyclists). In 1893, *Bicycling News* featured a front page article about how 'cycling mania' tends to 'exclude all else from some riders' thoughts' and which makes them 'lop-sided in their constitution.'[29]

Brands, tailors and companies identified cyclists as a new lucrative market. In particular, they targeted women. A plethora of instruction and advice was forthcoming on dress styles, materials, underclothing, shoes, veils, hats, hairpins, gloves, skin and hair care and much more. Many products purported to solve problems that women did not even realise they had. So motivating were these market forces that many existing products took on a cycling shape. Even new hairstyle fashions were not immune to the cycling craze. Advertisements for 'The Pneumatic Tube Coil' hairstyle promoted the 'latest novelty' for newly mobile women. Inspired by the invention of pneumatic inner tubes, these hair-wrapped coils apparently had the effect of making hair look fuller and did not get 'ruffled or out of order'. A 'clip on cycling fringe' was similarly marketed to newly mobile women. Examples like this provide a sense of how compelling new intersections of bodies, technology and freedoms of movement must have been at the time. Victorians could literally embody modernity. Even if you were not a cyclist, items like this meant you could still consume cycling, much like how sport brands are worn as street wear today. Together, these materials advocated a new enticing technological imaginary of and for mobile women.

Of course, the Lady Cyclist was a popular target not only for marketing but also debate and derision in the media throughout the 1890s. Cartoons, articles, fiction and poems on the topic guaranteed a readership eager to debate (and judge) what women should and shouldn't be doing on bicycles in public space. Fears of impropriety of a sexual nature were not far from these early cycling discussions. Sitting astride a saddle with legs moving independently on pedals raised concerns about the moral purity of the 'weaker sex'. Being outside, especially on one's own, beyond the structured control of family, male protectors and the home, could damage a woman's virtue. Although the attitude that cycling women were somehow 'fast' was mocked in satirical cartoons, there were some parts of society who saw it as 'an indolent and indecent practice that could even transport girls to prostitution'.[30] The belief that there was a link between a woman's brain and her reproductive system remained perniciously persistent.

The threat posed by cycling women, especially those attired in rational dress, arose from the perception that they were transgressing not only into masculine wardrobes but also privileged masculine spaces. As the editor of the *Daily Telegraph* noted: 'What with women voters, and women bicyclists in men's attire, old fashioned doctrines of women's place seem to be under going a process of rapid development'.[31] Cycling was viewed by some as interfering with women's 'natural role' at home, caring for others. There were fears of cycling leading to even more bad behaviour. A writer in the *Lady's Own Newspaper* mocks a claim made in other media that the 'middle class is smoking as unconstrainedly as the aristocracy, and the working woman is fast following' and that 'the bicycle is responsible for much, as with wheel parties has arisen a freedom of manner unknown in the presence of chaperones'.[32] Moral and gender instability confused people as to where an otherwise high-class woman could be located on the social spectrum, and some struggled with how to treat them.

Representations of women cyclists in the media similarly oscillated widely. Media opinion, as Clare Simpson notes, seemed to flip between dichotomous positions of women cyclists as either 'respectable or disreputable' and in doing so tried to 'make tidy an untidy situation'.[33] Historian Sheila Hanlon has argued that the 'lady cyclists portrayed in

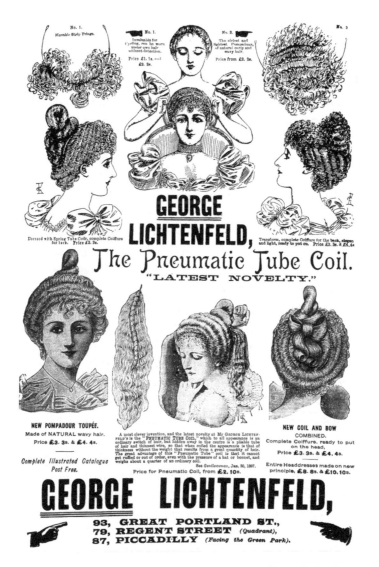

Figure 2.1 Advert for the 'Latest Novelty' Pneumatic Tube Coil, *The Queen*, 1897

comic images tend to fall into two categories: the naive butterfly rider unsteady on her machine and the manly scorcher with her bold mannerisms and bifurcated costume.'[34] Others, like Marilyn Constanzo in her study of *Punch*, claim that representations were not always trivialising and degrading. Skilled women were sometimes shown engaging in serious sports, which 'suggests that women challenged the dominant ideology simply by participating in sport and leisure activities alongside men'.[35] What is clear is that women on bicycles catalysed a spectrum of mixed reactions, which did little to define a path for early cyclists to follow. Instead, this ambiguity generated a new social landscape open to ideas, in which individuals could explore and experiment with possibilities, thus creating conditions ideal for problem solving and inventive thinking.

The Lady Cyclist was one of a slew of popular cycling periodicals that emerged at this time. Launched in 1896, although edited by a man, Charles P. Sisley, it featured many contributions from women, and provided much encouragement for women to cycle. Charles was supportive of many of the ideals underpinning the 'New Woman', arguing that 'women who are striving in various parts of the world to secure equal rights in political and governmental affairs for themselves, or their fellow-women should welcome the bicycle as one of the influences which will help them succeed in their efforts'.[36] He often responded directly to critics who habitually upheld the 'bad influences' of cycling for women, by calling them 'simply ignorant of the subject on which they are so fond of airing their opinions'.[37]

The magazine featured an enormous array of information to satisfy even the most cycling-crazed reader, from how to learn to cycle and what to ride, how to sit and not to slump, to where one should cycle, for how long and how fast. It covered popular cycling shows, such as the annual Stanley Show, in great detail and featured the newest design developments as well as practical advice on how to fix punctures, adjust handlebars, clean and grease parts, and the best tools to carry.[38] Being a Lady Cyclist in the 1890s was not limited to riding. She was encouraged to learn how the machine worked and undertake repairs when needed. What to wear was also of keen interest. Writers discussed what Lady Cyclists should and should not do in regular columns entitled 'Why a Lady Cyclist Should Always Dress Well', 'Cycling Fashions' and 'Mainly About Skirts'. There were illustrations of 'The Ideal Lady Cyclist' and columns about 'The Lady Cyclist at Home' as well as 'Which Are The Most Graceful Riders?' These articles, and many others like them, attempted to set the etiquette for cycling and cycle wear

Daughter (enthusiastically). " OH, MAMMA ! I *MUST* LEARN BICYCLING ! SO DELIGHTFUL TO GO AT SUCH A PACE ! "
Mamma (severely). " NO THANK YOU, MY DEAR ; YOU ARE *QUITE* ' FAST ' ENOUGH ALREADY ! "

Figure 2.2 The morals of a rationally dressed woman cyclist are questioned by an onlooker, *Punch*, 1895. Daughter (enthusiastically): 'Oh, Mamma! I must learn bicycling! So delightful to go at such a pace!'

Mamma (severely): 'No thankyou, my dear. You are *quite* "fast" enough already!'

and regulate it with many textual and visual examples of what did and did not work on and off the machine.

The cycling craze hit its peak in the mid 1890s and middle- and upper-class women were encouraged to take up this new leisure activity *provided* appropriate gendered decorum was maintained. The Lady Cyclist was encouraged, for example, to try to avoid becoming 'a woman who allows herself to be seen hot and red with exertion' as it was through this condition that she 'loses much of her feminine dignity'.[39] It was not only clothing that was regulated on a bicycle. Cycling bodies needed to be controlled too. Yet, cycling transformed bodies; they got hot, red and damp with sweat; skin became flushed; arms freckled, streaked with dirt and scraped by errant twigs; faces caught the sun and hair tangled in the breeze. Leaking unruly bodies were expected to be secret bodies, if at all. This is partly why swimming was considered an appropriate

Victorian activity for women (and similar beliefs continue today – for example, sportswomen's periods remain a taboo topic). Messy excreting bodies were often female – through pregnancy, menstruation, breast-feeding, crying – in comparison to, apparently, hard, reserved and con-tained male bodies. Posture was another regular target for critics, given a straight upright carriage was associated with morality. Not being able to properly manage your own body signaled a weak constitution, a danger-ous or rebellious character or an immoral streak.[40]

On the whole, publications like *The Lady Cyclist* and others advocated constraint and decorum and discouraged high-class women from show-ing effort or strain. Emphasis was placed on being 'graceful', 'neat', 'sensi-ble', 'becoming', 'charming', 'dignified' and 'modest' (note how different this is to the heroic discourse surrounding men and bicycles). It was deemed imperative that women appear feminine and exhibit, even while *mobile*, many of the desirable *immobile* characteristics of conventional upper-class femininity. Here for instance, Charles suggests a not-insubstantial list of things women should be doing to avoid negative public comments:

> Cycling can be both elegant and pleasing to the eye, but it can also be very grotesque. A woman must study every detail if she wishes to gain popular opinion and to look graceful both off and on her machine. Her costume must be simple, suitable, and neat, her machine kept in perfect order, and she must be able to ride in lady-like manner, and without the least apparent effort.[41]

With advice like this, it is not hard to see why some have argued that *The Lady Cyclist* fell far short of advocating any new political and gender positions because it reinforced many established feminine norms.[42] Paul Smethurst suggests that it is even possible to 'deduce that its main aim was actually the supression of radical views'.[43] Of course, what was written in the popular press and how women expe-rienced everyday life were not necessarily the same. However, what we can read into these various examples is the shift of values from the largely stationary body to the cycling one. Maintaining these strict social codes separated Lady Cyclists from others, like the 'New Woman' who often suffered spurious media attacks, as illustrated in this admonishment by the editor of *The Lady's Own Magazine*:

An exceedingly fast pace on a wheel ridden by a hump-backed caricature of the New Woman is a picture perfectly devoid of dignity and grace. There are, of course, times when exceptional pace is demanded, but this may be accomplished without assuming the ridiculous bend of the scorcher. A sensational gait or posture is always scrupulously avoided by the wheelwoman who would appear dignified and modest when awheel.[44]

Why did this happen? Why did social regulations that gave shape to middle- and upper-class women's lives, and limit them in many ways, get transferred to women who cycled? Some have argued that this was in fact a deliberate tactic to *encourage* women's cycling, and we can see evidence of this in the quote above. This seemingly counter-intuitive strategy assigned conventional respectability of middle- and upper-class society onto a particular form of women's cycling. Imbuing cycling with identifiable 'ladylike' characteristics was a way of making cycling a decent and socially accepted activity.

Claire Simpson offers another explanation of the contradiction at play: 'In order to convince their critics that, despite riding the bicycle, they were still feminine, some middle-class female cyclists tried to reconcile the ideology of the New Woman with conventional beliefs about femininity to create an alternative, yet still respectable, identity.'[45] This alternative identity fused familiar conventional gendered elements with new progressive ones, but stopped short of embracing all of the 'New Woman' ideals of equal freedom of movement. An example of this was provided by the editor of the *Bicycling News and Tricycling Gazette*, who took pains to point out that a woman could be a cyclist *and* not a threat to society: '[T]he lady who is a cyclist gives up none of her femininity, she is none the less to be domestically useful, she is not a "woman's rights" phenomenon.'[46]

This attitude helps to explain why many otherwise pro-women's cycling supporters expressed such vociferous opinions against women's racing. For them, racing was the antithesis of ladylike cycling; an extreme departure from feminine ideals. 'The Ladies Page' in *Bicycling News* lamented the 'cost' of ladies' racing that made everyone suffer the consequences: 'Those of us who remember how the same thing affected feminine cycling some years ago, and how long it took us to recover ground then lost, will feel that it is to the interest of all who have the real good of cycling at heart, to do all in our power towards discountenancing what has been, and is, so prejudicial to the pastime.'[47]

What is central to these arguments is the imperative of being recognised as a Victorian 'lady'. Cecil Willett Cunnington's *English Women's Clothing in the Nineteenth Century* provides insight into the nature of the shifting yet pervasive significance of this label. He writes about how the term 'lady' changed over time and 'became progressively more and more generous in its embrace, so ultimately it denoted not only an inheritance but an acquired status, even including some who earned their living'. However, 'it implied a special attitude of mind, of which correct conduct was the outward expression' and it 'was largely symbolised by her costume.'[48]

Clothing has long played a pivotal role in social ordering. Here, it was enrolled in the drive to imbue women's cycling with respectability. Provided a woman dressed like a lady and she held herself to some degree in line with recognised ladylike comportment, there was some flexibility tolerated with which she could press against deeply entrenched social norms and codes. However, the physical reality of cycling in ordinary clothing was no easy undertaking.

The Physical Reality of Cycling in Ordinary Dress

Popular periodicals were filled with cycling illustrations and studio portraits of Lady Cyclists posing on or near their bicycles.[49] These cycling women appear elegant, modest and charming. Their multi-layered fashionable costumes appear to fit well with this new mobility technology. This visual culture reinforced the idea that effortless cycling in ordinary fashion was possible. Skirts and hats remained in place, hair stayed unruffled and faces unblemished. These women cut a fashionable silhouette with their corseted waists and tailored coats. They are in control of their machines and their upright postures cut dashing figures in front of painted parkland scenes. These representations exemplified the ideal Lady Cyclist; a fusion of respectable Victorian femininity and demonstrable cultural cachet combined with (some) progressive principles (more on this in Chapter 9).

The reality, however, was considerably different. Wearing ordinary Victorian fashions on a mechanical object that moved at speed in varying environmental conditions could not be further from a staged studio portrait. Victorian streets were narrow and winding, paved with stone cobbles, wooden blocks or packed dirt, which turned to mud in inclement weather. They were crammed with horses and carts, omnibuses, new horseless carriages, darting pedestrians and stray dogs, and scattered with horse manure and rubbish. Cycling in these conditions was tricky. It was even more hazardous for women in ordinary dress. Even with skirt guards, short nose saddles and drop-frames, petticoats caught in chain rings, flew into spokes and wound around pedals. Skirts of a certain voluminous nature were even known to fly up and obscure the cyclist's view when coasting down hills or flap into companions' moving wheels. Newspapers regularly reported incidents of women being socially embarrassed, becoming terribly disfigured or even killed in bicycle crashes, with many journalists making specific reference to their clothing.

> I was in Landsdown (Cheltenham) one evening last week, and two young ladies were riding side by side, It was very windy, and the skirt of one blew into the wheel of the other, where it got caught. They both turned somersaults. When they were picked up, their skirts were very nearly taken off them – well, I found it necessary to look the other way. – *Daily Mail*, 1897

> I think she failed because she could not see the pedals, as the flapping skirt hid them from her view, and she had to fumble for them. The poor girl lingered a week. – *Daily Press*, 1897

> In London there was a very high wind throughout the day ... The most serious accident recorded was that which happened to Miss Agnes Ethel Hardington, of Chelsea, who on Sunday was blown over while cycling at St. John's-hill, Battersea and falling before the wheels of a passing milk-van was so seriously hurt that she died in a few minutes. – *St James*, 1899[50]

Figure 2.3 One of many familiar representations of lady cyclists achieving 'grace', 'modesty' and 'charm' on the bicycle, *The Cycling World Illustrated*, 1896

The deceased lost control of her machine while descending a hill, and was thrown to the ground. Her dress becoming entangled in the pedals, the unfortunately lady sustained injuries which proved fatal. – *The Rational Dress Gazette*, 1899[51]

This discourse reinforced the accepted notion of women as the 'weaker sex'. They were technologically incompetent. They lacked confidence and skills. They were not good cyclists. This is despite the fact that many of their crashes were caused by their clothing. Kitty's brother, S. S. Buckman, was a regular contributor to *The Lady's Own Newspaper* writing under the pseudonym 'A Wheeler' in a column 'Wayside Jottings'. He often put forward a different opinion, arguing that '[t]hese mischances are the experience of everyone riding in skirts'. He offered an alternative reading of these grim stories: '[T]hat more serious accidents do not occur often is ... only due to the fact that those who ride in skirts are so handicapped with weight of machine and wind resistance that they cannot, and do not, ride at any pace.'[52] It could also have been the case that women were more skilled and careful riders as a result of having to adapt to these technical and material impairments.

The Social Dangers of Cycle Wear

Although many cyclists were aware of the dangers of wearing ordinary dress on a moving bicycle, the choice to wear more appropriate or rational cycling attire was not necessarily safer. Women felt the social impact of their sartorial choice not only in vigorous media debates, but also personally on the streets. Why was the bifurcated garment for cycling in particular so provocative? Women had been adapting masculine clothing in other sports, such as horse riding with its tailored jacket, silk hat and cravat, and even Queen Victoria had worn a version of a men's military jacket and cap in the early nineteenth century. According to Diana Crane, trousers catalysed such ferocity because they were so firmly linked 'with male authority'.[53] Knickerbockers and bloomers might have kept skirts out of the wheels and women safer from cycling incidents but they exposed them to new dangers in the form of verbal and physical abuse on a daily basis. On occasion it was so violent that women's cycling companions were shocked at how the public treated them:

> What females who adopt the semi-masculine costume have really to put up with I had no idea till the other night, when as I was walking home, I was passed by two girls who were thus attired, and they were being assailed by such a torrent of foul and obscene language ...

I am sure that no man, however much he might be in favour of the 'new woman attire', would ever submit any female for whom he had the slightest regard to the change of such a disgusting attack.[54]

Rational dress remained deeply divisive. The editor of 'The Ladies Page' in *Bicycling News* provided this advice to a fledgling new cyclist: 'An ordinary walking skirt is by far the best for all-round wear, since it looks well in the saddle, and out of it, and it does not stamp "cyclist" upon its wearer'.[55] The Rational Dress League regularly voiced concerns about the unacceptable nature of abuse that women were facing in public and how this was shaping their clothing decisions:

> We are told by The West End that the crux of the matter, i.e., the wearing, or not wearing of rational dress, depends not on what women wish to wear but upon what the cab drivers, the omnibus conductors, the corner loafers and the street boys *will permit them to do*. But this is perfectly monstrous.[56]

For those who persevered the social stigma could be harsh, especially when men in powerful positions took matters into their own hands. An article in *Bicycling News* under the heading 'Dress Reform in Peril' tells of the Prefect of Police's personal distaste of rational dress. He apparently was so concerned that he 'put in force a law against women masquerading in male attire' and which had already 'been served upon several women cyclists, more as reminder that they must abandon the practice than with any intention of further proceeding against them'.[57] While not actually legal, and a clear abuse of power, his aim was clear – to prevent the uptake of rational dress before it took hold. A similar attitude was still in place nearer the turn of the century. The hostile attitude of road users towards women cyclists became so pervasive that it was raised in a Cyclist's Touring Club meeting in 1898.

> A new and important point is the decision to take legal action against drivers who willfully endanger the lives of 'rational' cyclists and against any persons using insulting or obscene language at rational dress wearers. This is in a field in which, it is thought, much may be done to show the loafer and the street-urchin that they have

no more right to use foul language to a lady in knickerbockers than to a lady in a skirt.[58]

That same year, Richard Cook of the White Horse Inn, Dorking, announced he would not allow women cyclists attired in rationals to enter his establishment. It was an attitude supported by P. Maurice's 'Letter to the Editor' in the *Daily Mail*:

> To the proprietor of the White Horse Hotel, Dorking, earns the thanks of all lady cyclists for the plucky stand he has taken. We now know where we can take our sisters and other people's sisters without fear of being sickened by the sight of these 'middle-sex' women, who are neither true ladies nor true gentlemen. Three cheers for him.[59]

To people like P. Maurice, rationally attired women were viewed as 'sexless', having given up their gender and all the rights this gave them in Victorian society, as a result of seemingly having rejected the domestic sphere and associated wifely or daughterly duties. The infamous incident of Lady Harberton being denied entry to the coffee room in the Hautboy Inn, Surrey, came not long after this. This altercation was on account of her wearing rational dress, and it drew even more media coverage, public opinion and a court case.[60] In a society fiercely defined by gendered roles and responsibilities, these actions had deeply unsettling and upsetting social ramifications. Many viewed these women as not only entering into established masculine territories but of threatening to entirely transform gender relations.

Even as the end of the century loomed, Victorian society was no closer to reaching consensus on an appropriate cycle garment for women. An illustrative example of the extent that women's cycling and cycle wear *still* represented a threat to social norms came in response to a proposal put before Cambridge University's Senate to grant full equal degrees to female graduates. Male students responded with a protest that featured an effigy not just of a female student but a woman on a bicycle dressed in rational attire. The dramatic hanging of this effigy from a college window represented furious disdain to women's multiple claims to public space – education, cycling and unchaperoned independence. As reported in *The Cheltenham Examiner* on 2

Figure 2.4 Box O'Lights My Lord? A rationally dressed lady cyclist is mistaken for a man by a small boy and people laugh nearby, *CTC Gazette*, 1897

June 1897: 'The effigy of a lady clad in bloomers upon a bicycle was suspended opposite the Senate House, causing much laughter.' The female cyclist was apparently clad in 'blue trousers, pink bodice, large goggles, boating shoes, yellow and mauve striped stockings, and an old college cap'. Emotions were raised even higher in the aftermath of the protest when news circulated that the resolution did not pass. Sheila Hanlon writes about how 'the triumphant mob tore down the effigy' and 'savagely attacked the mannequin, decapitating and tearing it to pieces in a frenzy'. Afterwards, '[w]hat remained of the poor lady cyclist was stuffed through the gates of Newnham College, the nearby women's college'.[61]

Creating the Conditions for Invention

Women's clothed bodies did not initially fit socially or physically with the bicycle in Victorian society. Enthusiastic cyclists sought many ways of making them fit. Although many were aware of the dangers of wearing conventional ordinary dress on a moving bicycle, the choice to wear more appropriate cycling attire was not necessarily safer. As illustrated by Kitty and her friends, to cycle in more rational attire, such as bloomers and knickerbockers, could result in verbal abuse, in some cases physical assault and sometimes more insidious social stigma. Women were subjected to volleys of rocks, sticks and rude remarks. They were

RATIONAL COSTUME.

The Vicar of St. Winifred-in-the-Wold (to fair Bicyclists). It is customary for Men, I will not say Gentlemen, to remove their Hats on entering a Church!"
Confusion of the Ladies Rota and Ixiona Bykewell.

Figure 2.5 Rational Costume – The Vicar of St. Winifred-in-the-Wold (to fair Cyclists): It is customary for men, I will not say Gentlemen, to remove their hats on entering a church, *Punch*, 1896

gossiped about and denied entry to public places. Some viewed them as sexless, having rejected the roles and responsibilities of their home and gender. In the process of moving beyond defined gender roles, through clothes and actions, they were forging new ways of being in the world, as differently mobile and gendered indivduals, and many suffered the consequences.

Remarkably, little appeared to dampen their enthusiasm for cycling. Once experienced, the unparalleled freedoms of the bicycle could not be easily forgotten or ignored. Some saw it as a political opportunity to carve out new identities between conventional binaries of reputable and disreputable womanhood. Others saw it as a platform to campaign for women's rights or as a business opportunity. Some just wanted to ride their bicycles. What these conditions created was the impetus for many to take matters in their own hands – to do it themselves. Given clothing's high status in society, it was not just a practical but also political place to start. This required a lot of work on many fronts. Some had to deal with personal issues, at home and in familial relationships. Others had to come to terms with the level of distress they could tolerate in public. All had to mobilise skills, networks and resources at their disposal. Every experience was different. And women rose to the challenge in a spectrum of creative ways.

3

Inventing Solutions to the 'Dress Problem'

Men can never know, unless they try the experiment themselves, how heavily a wheelwoman is handicapped by her skirt. Even without wind the loss of power through friction is immense, while with it the pleasure of riding is often changed to misery. The extra exertion required to propel the machine, when your skirt is acting as a sail to blow you back, is bad enough even without the consciousness of looking untidy and ungraceful.[1]

Ada Earland
Dress Reform For Women, Bicycling News, 1893

The 'dress problem' for women was two-fold – how to cycle safely and comfortably while maintaining charm and decorum. In reality it was difficult to cycle in ordinary dress. It became untidy easily, slowed a woman down and had the potential for danger when it caught the wind like a sail and flew into moving parts. According to Ada Earland, ordinary dress 'handicapped' a woman and caused 'misery'. Although more rationally oriented cycle wear enabled women to travel further and longer, in more comfort and safety than before, it was not universally adopted. In a society founded on highly regulated rules and traditions, it could single a woman out as embodying a political act and she could become a target for condescension or worse. While men could for the most part relish their public cycling identities, women were made acutely aware of the impact their mobile presence had on others. Although this was undoubtedly distressing at times, these conditions created a new landscape of social, material and technological possibility. This chapter tells stories of some of the creative and inventive responses to the 'dress problem'. What emerges is that there was no *one* solution. There were many, and

some of these inventive responses continue to shape how we cycle over
a century later.

Bicycle Design Strategies

One of the more familiar creative responses to women's dress problems emerged in relation to the design of the bicycle. Throughout the
nineteenth century, the bicycle had undergone a dramatic and accelerated technological trajectory, from the Dandy Horse (an assisted
running machine) to the Bone Shaker (with front-wheel drive) and
from the Ordinary or 'Penny Farthing' (with increased front wheel
to maximise speeds) to the Safety (a diamond frame with rear-wheel
drive that we are familiar with today). With each of these designs, and
others in-between, the mechanics and nature of cycling was radically changed. Improved gear ratios, pedaling systems and pneumatic
tyres meant people could travel faster over greater distances and
with increased ease. Women's dress, however, remained consistently
problematic.

Layers of flapping materials and moving mechanical parts do
not mix well. With the Ordinary tricycle, women sat on a platform
seat located between the large wheels. Skirts and petticoats gathered
and rose problematically up the rider's legs when she pedalled and
tended to blow into pedals and spokes. While women were keen tricy-
clists, riding was an expensive and, as a result, exclusive hobby. This
changed with the advent of the Safety bicycle in the late 1880s and
women's enthusiastic adoption of this form of cycling helped drive
the cycling boom of the 1890s. This smaller-wheeled bicycle was more
affordable than the high-wheeler (and became increasingly more so
over the course of the decade), as well as being lighter and easier to
store, maintain and navigate in busy city streets. Initially, Safety bicycles were diamond shaped with a top bar that required the rider to
mount their machine by lifting a leg up over the frame. This was physically taxing in heavy layered garments, often inelegant and potentially
embarrassing, not to mention tricky, for women in petticoats and long
skirts. Recognising the potential of a new market of consumers, bike
manufacturers began to expand and diversify their catalogue.

Cycle makers set out to adapt the Safety bicycle to fit women cyclists, paying heed to the social and material restrictions at play. There were various iterations but ostensibly new designs worked with the status quo; they materialised accepted ideas about 'grace', 'modesty' and 'charm' and attempted to work around the problems of conventional dress. Manufacturers accommodated women's skirts by removing the top tube of the diamond frame and creating a drop-frame. These bicycles featured short-nosed wider saddles, dress-guards on the rear wheel and chain covers. A rider could step through rather than over the frame and the extra space made room for bulky skirts. She could sit upright and cycle at a pace more apparently suited to her gender, and avoid unnecessary (and visible) exertion.

The drop-frame bicycle soon became known as the 'Ladies' Bicycle'. The availability of this customised Safety carved out a space for women in the male-dominated cycling world. Lady Cyclists started to appear in advertising campaigns. They were directly targeted as consumers and a new visual language placed women firmly in cycling's fashionable set, giving many permission to continue their sport and providing incentives for the tentative. However, while designs enabled many to take up cycling, and to do so in ordinary fashions, they were not without their critics. Some felt this design solution did not directly address the 'dress problem', as skirts could still entangle dangerously in the wheels and pedals. Frustrations flared at specific times, particularly around popular bicycle events such as the Stanley Show, where manufacturers exhibited cutting-edge products. As women were quick to note, inventions tended to focus on men's needs. They wanted to see similar advancements in bicycles that *they* could ride. Marguerite, a regular writer of 'Lines for Ladies' in *Bicycling News and Sport and Play*, regularly lamented the limited range of bicycle designs available for women:

> What a pity it is that manufacturers do not, as a rule, bestow the same amount of pains on the production of a machine for ladies' use as they do upon those intended for the use of the stronger and sterner sex. The very extremes of lightness are reached, so far as men's machines are concerned, men of twelve stone riding machines weighing only about twice as many pounds; whilst a woman, who is

lighter as well as weaker, and who is usually more careful as regards the usage of her machine, is obliged to be content with one scaling ten pounds heavier ... the majority of ladies' machines weighing considerably more than thirty pounds.[2]

Ironically, even without the extra top tube, women's frames tended to be significantly heavier than similar men's machines. In 1895 a writer in *The Queen* optimistically assumed that '[t]he rapidly increasing demand for ladies' bicycles will doubtless encourage makers to devote more attention to the very important question of weight.'[3] Lighter machines would have made cycling easier for women, but innovation required not only recognition of women cyclists as a viable consumer group but also as drivers of technological change. But, the larger consumer market tended to follow racing trends, and racing was almost entirely male. Competition between manufacturers to build cutting-edge men's machines mirrored rivalry on the racing track. Everyday male cyclists benefited from this because, as writers noted, 'the racing machine of one season often becomes the light roadster of the next.'[4] This dynamic market encouraged makers 'to gradually reduce weight by a careful study of the more minute details of construction' so that men's machines weighed only 20lbs (approximately 9kg). This is important because it further associated men, even those who were everyday casual cyclists, with high-quality, professional and cutting-edge technology which enabled them to assume the role of technically confident consumers. This position was not available to women then, and it seems even now, given that women continue to report discrimination in many cycle-oriented contexts.[5]

In 1896, *The Hub* reports that women's wheels 'have been brought down from 50 to 24lbs, or under within a comparatively short time', however, they remained 'somewhat less rigid and strong than men's wheels.'[6] This is a massive shift, from almost 23kg to 11kg (and it prompts reflection on how weak the 'weaker sex' was to be able to manage a bicycle of that initial weight, while wearing multi-layered bulky clothing). In addition to the disproportionate weight of women's velocipedes, the range of sizes were limited in comparison to the male cycling market. As Marguerite points out, one size did not fit all women.[7]

It would be only reasonable, however, on the part of manufacturers, to make machines for ladies as they do for men, in at least two sizes, when a tall girl would no longer have to ride in such a huddled-up position, nor would she be forced the only alternative which has hitherto been open to her of riding with handle-bar and saddle-pillar extended in a manner which not only spoils the symmetrical appearance of the machine, but is also a source of danger to herself.[8]

This is particularly important, as women were regularly criticised for not sitting or pedalling properly, and this was often used as evidence of women not being 'good' or 'natural' cyclists. As contemporary cyclists know, it is a difficult thing to do when the bicycle is the wrong size. Observations such as this, in *The Queen*, were common:

Everywhere the number of lady cyclists continues to increase, but it is a regrettable fact that the really graceful and efficient rider still remains a pleasing novelty. The majority of riders that one sees are sitting so low that their knees come up at right angles to their bodies, their insteps are on the pedals, and, to add to the general hunched-up appearance, they have their handles about six inches too high; the wonder is that they get along as well as they do, and it only shows what remarkably good riders most women would be if they only took a little trouble with their form.[9]

Moreover, the drop-frame design and dress-guards were not universally welcomed. S. S. Buckman, again writing under the pseudonym 'A Wheeler' in *The Lady's Own Magazine* conveyed his low opinion of the design by declaring: 'An open frame is only an apology for a bicycle.'[10] Others argued that these design solutions compromised the strength of frames, and the additional features added further unnecessary weight and fiddly components that caught flapping materials. Many women, understandably, were against the idea of adding even more bulk to their bicycles, especially if they were persisting with skirts and petticoats. Some had extra parts removed; an act that was criticised by many, such as this journalist from *Harper's Bazaar*: 'It seems to me that any woman who wears skirts when bicycling is reckless in removing her chain-guard.'[11] Cynthia dedicated an entire page of her regular 'Little Essay's by Cynthia' in *The Lady Cyclist* to dress-guards, which she thought could 'scarcely claim to be an indispensible part of a bicycle'. She imagined a future in which 'the dress-guard will only be in a cycling museum, along with the boneshaker, the rear-steering, single driving tricycle, and specimen cycling costumes of the years '95 and '96, with skirts five yards round'. Cynthia believed the dress-guard was only tolerated by skirted riders and then was not always reliable. She declared that 'we have already consigned our dress-guards to the "chamber of horrors"'.[12] Others directed their frustrations not at the dress-guard but at the skirt. Marguerite of *Bicycling News* was adamant that 'in most cases the skirts of the riders rather than the inefficiency of the dress-guards, are conducive to accident'.[13]

What is clear is that the Ladies' Safety, and various accountrements, did not solve the 'dress problem'. They worked with and around existing issues which many women found unsatisfactory. Some even refused outright to ride women's specific bicycles. Mrs Smith was one such cyclist who sent her illustrated testimony to 'The Ladies Page' in *Bicycling News*, which prompted the editor to comment:

> I must thank Mrs Smith who has just kindly sent me two beautiful photographs of herself in the rational dress in which she appeared a week to two ago at Ditton. In the photographs Mrs Smith is riding a man's safety, her dress enabling her to overcome the difficulties of the top bar with complete ease. It is one argument the more in favour of rational cycling costume, that with it a woman may dispense with the drop frame, which always means a loss of strength to the machine.[14]

A similar view was voiced by Ida Trafford-Bell in the *New York Times* about pairing her rational dress with a diamond frame machine. 'Of course, the knickerbocker and bloomer costumes go with the diamond frame or man's wheel, which all sensible men and women are forced to admit is the only correct and proper machine and dress for long rides, touring, etc.'[15] Mrs Hudders went one step further. She was a keen bloomer-wearer and adamant that riding a diamond frame was the only 'wheel' worth riding:

> And if I have to chose between giving up my wheel and giving up my diamond frame, I should give up the wheel altogether. You can get

a better position on a diamond frame, do less work riding, there is less vibration, and there is greater safety. The wheel is lighter, too.[16]

These accounts reveal how debates around bicycle designs for women were never separate from dress. There was even talk at the time that the Ladies' Safety frame would only be a temporary solution. A journalist in the *New York Times* writes: 'A well-known manufacturer gave it as his opinion recently that there will be no women's wheels made in three or four years, for all women riders will be wearing bloomers or knickerbockers or something like either or both, and will be able to mount their wheels as men do.'[17] A. Wheeler (S. S. Buckman) conveyed similar attitudes of manufacturers in Britain: 'It's very sad – and more sad still to think, as my husband tells me – that behind the ladies backs the manufacturers grumble about ladies fads, about stupid drop frames, and about fiddling dress guards, and ask why ladies can't dress so as to ride bar-frames.' He suggests that bike-makers actually wanted to build a much more suitable bike for women if they could just get beyond the dress problem. 'Said a manufacturer a while back, "There's a lady as light as a feather, how I should like to build her a little light bar-frame: she'd regularly fly on it. I could do it for £2 less money, and I could make it 5lbs. less weight, and stronger than the drop frame that I'm to build her ever will, or can be".'[18] The problem, he thought, was not just the vagaries of fashion and buyers but also distributors who had built up a new market around solving women's 'dress problem' in this way and had to clear stock.

> The adoption of Rational dress should mean the adoption of the bar-frame machine, and that would mean an enormous depreciation in ladies' drop-frame cycles, and the parts thereof. Now there are thousands and thousands of pounds worth of drop-frame machines made up, and an equally great amount of fittings which can only be used in the construction of such machines. A change in fashion would be disastrous to those who hold this stock.[19]

Cycling manufacturers were quick to pick up on fashions to sell bicycles, partly to overcome the problem of people not needing a new machine once they owned one. As a result, they were sensitive to fashions and social whims which might explain the lack of attention on women's specific machines. If Ladies' Safety designs emerged in response to the 'dress problem', and this impediment was considered temporary, then the need for drop-frame velocipedes might cease to be relevant when fashion inevitably changed and women took to new forms of dress. The fact that some big brands like Triumph advertised their women's veloci-pede as 'A Fashionable Machine' points to this. A. Wheeler seemed convinced: 'Hence much of the opposition from the wheel organs – opposition that will be swept away by the popular taste, at will most in a couple more season.'[20]

Many cyclists were also aware of changing fashions abroad. In her wonderfully named column 'Array Yourself Becomingly' in *The Cycling World Illustrated*, Virgina laments the drop-frame style for women's cycling: 'I suppose you know that in Paris it is almost impos-sible to buy a woman's machine, at least one made on the same line as ours – the cross bar is ubiquitous, although the make of the machine adapted to feminine use is somewhat lighter than that which is devoted to mere masculine service.'[21] Maybe designers did not think that continuing to innovate in this field was a valuable investment of time and resources. Change would have felt constant for Victorians. It is useful to remember the accelerated trajectory of the bicycle and radical reform to dress in their lived experience. As this and the next chapter attests, late nineteenth-century society was buzzing with the excitement of inventiveness. Even small mundane ideas held the promise of sweeping change.

While the Ladies' Safety played a critical role in Victorian cycling cultures, it was not a straightforward or universally agreed solution to the 'dress problem'. It is easy to assume, given the fact that the Victorian ladies' frame is still with us today over a century later, that it had a smooth and uncontested technology trajectory. It didn't. The design of ladies' bicycles was in some places quite sceptically received. While some favourably adopted it, others did so grudgingly. Some com-plained and attempted to modify it to fit their needs. Others snubbed it entirely and chose to ride a diamond frame. For many, including some manufacturers, it was apparently viewed as a temporary fix to the immediate 'dress problem' of lady cyclists who wanted to ride in skirts. Specifically targeting women was also a way to grow the cycling market.

THE
LADY
"TRIUMPH."

The Fashionable Machine

FOR

1897.

TRIUMPH CYCLE CO., LTD., COVENTRY.

48, HOLBORN VIADUCT, LONDON, E.C.   26, SHAFTESBURY AVENUE, LONDON, W.

Figure 3.1 Adverts for women-specific bicycles, *The Lady Cyclist*, 1896

These debates are still important because segregated designs continue to shape and influence today's cycling identities and behaviours. While both women and men rode the Ordinary tricycle in the 1880s, and a limited number of models such as the 1876 Starley Coventry Lever Tricycle featured some adaptation for women's dresses, it was the advent of the Safety that directly shaped the bicycle we know today into two distinctly gendered objects – a men's diamond top-bar frame and a ladies' drop-frame. Yet, as discussed, more innovative energy was poured into the development of men's machines as a result of the direct link with racing culture, which was dominated by men. This imbalance is in many ways still evident today. No doubt it would have come as a surprise to many Victorian cyclists to have known that versions of the Ladies' Safety, with many of its attending problems of weight, lack of range and technical specificity, remain with us even though the design assumptions born of the social and material conditions of its origin do not. Even stranger is the fact that British women and men now wear similar clothes to cycle that fit their bodies to all kinds of bicycles, and yet gendered bicycle designs still persist.

Cycle Wear Strategies

Another strategy for addressing the 'dress problem' was more direct. Attention was focused on clothing as a means of mediating the relationship between women's bodies, society and the bicycle. Some designers sought to fix the skirt itself, and in the process many found themselves reconfiguring social parameters along the way. For others, social change was the primary aim and cycle wear was a means to attain it. Regardless of intent, there were a number of tactics involved. What follows are not separate approaches but rather a flexible and dynamic spectrum of design responses, some of which overlap, and many interconnect. Critically, what emerges is how women were constantly and creatively responding to changing conditions through imaginative combinations of bodies, clothes, networks, spatial practices, skills and courage.

Rational Dress

Cycling in identifiable rational dress is perhaps the most commonly known sartorial cycling strategy of this period. This involved replacing cumbersome petticoats and skirts with bloomers or knickerbockers (though there were also variations with skirts of differing lengths and discussions around the wearing of loose or no corsets). While the Dress Reform Movement had been around for most of the century, the popularity of sport and particularly cycling propelled it into the public domain, generating the public attention it needed. While the wearing of rational dress rarely made for seamless social encounters, there is significant evidence that many women did it. Kitty and her companions regularly wore their rationals in public and shared experiences of reactions to their costumes.[22] Their letters describe how they dealt with the emotional and material labour involved. One needed 'nerves of iron' not only to bear the vitriol but also to deal with the dramatic shifts in observers' responses from place to place:

> We stayed the night at the 'Katherine Wheel' Shrewton, kept by pleasant people who much approved of Rationals and we were very comfortable there and cost us 9/.[23]

> Two girls on bikes passed me one day and one shouted 'You ought to be ashamed of yourself'. I was ashamed of her and her lack of manners.[24]

> There were cries of 'Bloomers!' 'Take em off' (so idiotic that) but nothing to hurt.[25]

Some rational dress wearers were less fortunate. Occasionally reactions from the public did hurt. Writing in 1899, Irene Marshall's account illustrates how grim it must have been at times to attempt to claim a rationally dressed female cycling identity:

> But it took some courage five years ago to ride in rationals. The idea was almost entirely new and the British Public was dead against it. Hooting and screeching were then the usual accompaniments to every ride. Caps, stones, road refuse – anything was then flung at the hapless woman who dared to reveal the secret that she had two legs. And the insults were not confined to the lower classes. Well-dressed people, people who would be classed as ladies and gentlemen, frequently stopped and made rude remarks. In fact, cycling in rationals in 1894 was a very painful experience.[26]

Lady Harberton recognised the importance of this (uncomfortable) attention for the Rational Dress Movement and was famous for not shying away from social strife. In personal correspondence to S. S. Buckman she writes: 'On Sat I was in Tewkesbury, I noticed my appearance created much disapproval, so, if fine I am going there again tomorrow for tea.'[27] She led by example and urged women to put their new radically clothed bodies on display in public to claim the right to cycle independently, safely and comfortably. Although she knew this was not going to be easy for everyone, nevertheless she advocated wearing the costume at all possible opportunities, on and off the bike, in an attempt to normalise it through familiarity and ubiquity. She explains: 'In the first place, while no absolute rules have been laid down to bind members, we naturally expect that those who earnestly desire freedom in dress will do all in their power to help; not only by donning the costume for special meetings and rides, but by wearing it on every possible occasion.'[28]

Despite the breadth of coverage and debate generated by the Rational Dress Movement and positive reactions of those who embraced it, it was not considered a success. Lady Harberton expressed her disappointment in the movement in personal correspondence to S. S. Buckman on 18 April 1898: 'Do you know I don't much believe in a conversation on anything doing much good except people choosing to wear the dress. And how to get them to do this is the problem.'[29] However, Lady Harberton was not easily disheartened. Like other women at this time, she determinedly took to the next newest form of mobility technology that promised emancipation – motoring – as soon as she could and helped to drive it into the new century.[30] Lady Harberton was committed to new technologies and women's freedom of movement in many forms.

Strategies of Concealment

In trying conditions, some women were understandably hesitant to wear a rational costume in public. A related strategy involved wearing a version of it in less overt ways. Replacing petticoats with knickerbockers or bloomers was one tactic for preventing some of the mishaps caused by layers of heavy flapping fabric. With such radical undergarments hidden *under* skirts, the potential for social friction was also lessened. Marguerite of *Bicycling News* advocates this concealment strategy:

> I presume there are few ladies who ride in the old-fashioned under garments: I sincerely hope there are *very* few, as they give a most ungraceful appearance to the rider, and to ride in comfort with a number of petticoats to which some women so fondly adhere, is utterly out of the question. I know there are many who would not care to don the 'rational' costume for riding, and whatever its votaries may say to the contrary I can assure my readers, from personal experience, that it is quite possible to ride in a skirt with complete comfort and immunity from danger providing the under garments be right.[31]

More dedicated rational dress supporters were less taken by this tactic. After all, their larger strategy was to claim the right for women to move in new ways in public space. Concealing progressive clothing *under* conventional skirts defeated this purpose. Kitty's cycling companions often discussed how they tried to encourage each other to wear rational garments in public. Jane writes how 'Minnie wouldn't let me ride in a skirt around Andover and as she did not herself, it seemed foolish for one and not the other to do it, but I hate skirts more and more and ride in one as little as possible.'[32]

Many believed, like Minnie, that it was essential that rationally dressed women claim outside space in rational dress en masse. Yet, it was a topic vigorously debated within dress reform circles. While the Rational Dress League on the most part welcomed 'ordinary dress wearers' in the hope for 'gradual education of the public,'[33] members of the Lady's Cyclists Association were more conflicted. They vigorously debated in their 1896 annual meeting if rational dressed riders should be allowed onto 'skirted' rides when the reverse was not the case.[34] Nevertheless strategies of concealment remained a popular strategy in many cycling worlds and might well have been an initial step for newly minted riders to try out rational garments en route to becoming more fully fledged members.

Site-Specific Cycle Wear

Another strategy involved site-specific cycle wear. Some women responded to social and material challenges by adapting their costumes according to place, type of cycling and anticipated public context. This meant choosing a conventional fashion garment for social and public forms of cycling, at an invariably slower pace, and a more appropriate garment for 'real' or 'actual' out-of-town cycling or touring. This was not a secret strategy. Writers espoused this option for lady cyclists. '[C]ycling dress for town and for country is quite a different thing', declared Miss F. J. Erskine in her 1897 book *Lady Cycling: What to Wear and How to Ride*. Similarly, Mrs Selwyn F. Edge, in an interview for the regular column 'Lady Cyclists at Home' in *The Lady Cyclist* confirmed her site-specific strategy when asked:

> 'What costume do you wear?'
> 'A skirt for town riding, but rationals for actual riding.'
> 'Which costume to you prefer?'
> 'I certainly think a skirt looks best when riding slowly, but for real riding I like the rational. The skirt flops about so terribly when riding quickly.'[35]

Courting was another motive for this site-specific approach. Cycling offered new exciting opportunities to meet, spend time and interact with potential partners. It was therefore important to invest time and effort into maintaining appearances because 'courting demanded the most attractive clothes one owned.'[36] The *Lady's Own Magazine* confirms this: 'The greatest matchmaker of the age, to my mind is the bicycle' because 'young people are brought together by a common interest in wheeling, and this companionship often leads to an alliance for life.'[37] For many, engaging in these rituals continued to perpetuate socially accepted ideals of feminine grace and effortlessness, even while mobile.

Country-Specific Cycle Wear

Site-specific strategies were translated more broadly into what could and could not be worn in different countries. Cycle touring was a popular pastime for Victorian women and many published excerpts from their travel diaries. They had experiences of seeing and participating in cycling cultures abroad and returned to Britain freshly attentive to the possibility of other ways of moving in and through public space, and shared their experiences widely. *The Lady Cyclist, The Rational Dress Gazette* and *Harper's Monthly Magazine*[38] all offered condensed narratives of touring at home and on the continent. Many tourers enjoyed the freedoms of other cultures and lamented having to return to more conventional, restrictive costumes and ladies' bicycles in Britain. A contributor to *The Rational Dress Gazette* remarked: 'The tolerance which prevails in France in respect to feminine rational cycling dress, strikingly contrasts with the intolerance displayed towards it in England.'[39]

An illustrative example is provided by Fanny Bullock Workman's *Notes of a Tour in Spain*. She recounts her experience of holidaying in 1896 with her husband, two bicycles and a trunk of clothing. The latter was sent in advance to destinations en route. After catching a train to Tarascon in France, the prose combines cycling adventures and cultural insights with woes of administrative customs and visa delays encountered along the way. Fanny owned a rational costume, comprising knickerbockers or bloomers and a tailored jacket. She had done away with the skirt entirely. She confesses that this is the first time she had the courage to wear it and is pleasantly surprised by attitudes to women's cycling in France, in contrast to public opinion at home. She was also amazed at how quickly normalised it became.

> In France, the birthplace of the 'rational' dress, I had decided to don mine for the first time, and it was with some trepidation on the morning of our start from Tarascon that I approached my machine, which was surrounded by a crowd of admiring townspeople. But they never gave me a glance, and after two days of riding I should have been ashamed to have been seen in any other costume, for besides the delightful freedom it affords in riding, I saw nothing else worn by women, even in the south of France.[40]

Given experiences like this, it is no wonder travellers attempted to replicate the freedoms they had enjoyed on holidays when they returned home. In most cases, this was not an easy transition. Written in 1894, Lady Violet Greville's *Ladies in the Field: Sketches of Sport* explains how

Figure 3.2 Love's vehicle in three centuries, *The Lady Cyclist*, 1896

women often tried to follow the example set by their European com-
patriots: '[A] few Englishwomen have appeared on the public roads in
knickerbockers, and have made, as was to be expected, great talk in the
cycling press.'[41] Although it was only 'talk' and not sticks or stones, the
fear of social stigma and its ramifications would invariably have served
to keep this kind of behaviour in check.

Making (and Patenting) Your Own Cycle Wear

This strategy, and the focus of the rest of the book, involves the design
and making of new forms of cycle wear that sought to directly address
many of the restrictions and challenges facing newly mobile women.
This involved inventively responding to the 'dress problem' in and
through the dress itself. Although a ready-to-wear market was emerging,
many women continued to make their own costumes or commission
someone to do it for them. There are crossovers, of course, with strategies mentioned above, as many women were probably making their
own rational dress and wearing it in specific places. However, the lack

"*Asked the way to the road that led to Gerona.*"

Figure 3.3 Fanny Bullock Workman, dressed in her rational cycling costume, asks for directions in Spain, *The Lady Cyclist*, 1896

of a single socially agreed cycling costume for women catalysed many individual responses. The decision to patent designs also opened up for women a broader landscape of social, cultural and political possibility. It brings to light the many skills, networks and technologies women mobilised to enact change.

Although focused on the history of embroidery, Rosika Parker's influential book *The Subversive Stitch* is relevant here in terms of changing ideas about feminine ideals and expectations. From the seventeenth century, feminine behaviour became aligned with ideas around class, immobility and the home. Embroidery was associated with 'stereotypes of femininity', about 'docility, obedience, love of home, and a life without work – it showed the embroiderer to be a deserving, worthy wife and mother.'[42] However, it was also more than this. Parker argues that women did not passively learn these skills and accept these beliefs but used them to rework and negotiate their place in society. She writes: 'The art of embroidery has been the means of educating women into the feminine ideal, and of proving that they have attained it, but it has also provided a weapon of resistance to the constraints of femininity.'[43]

These ideas can be applied more broadly to sewing. Like embroidery, many women are acculturated into sewing at an early age. It is a multi-faceted practice for many classes, undertaken for utilitarian and leisure purposes, and passed down through generations. It has also enabled women to engage in work from home. In her book *'Make it Yourself': Home Sewing, Gender and Culture 1890–1930*, Sarah Gordon writes about how 'sewing represented the home, women's conventional role of caring for her family, and was associated with the concepts of thrift, discipline, domestic production, even sexual morality.'[44] For upper-class women needlework formed part of their cultural education, along with an understanding of music, art, literature and the ability to dance. Despite its prevalence, it is remarkably understudied. Yet, it has provided long-lasting artefacts that 'stand as evidence of women and women's activities.'[45] Material of this kind is valuable, especially when women's voices and contributions are otherwise absent in historical accounts.

Home-made cycle wear is evidence of early women cyclists using the tools, skills and networks available to them to redefine the boundaries that restricted them. Many attempted to sew a way out of the 'dress problem'; to find a way to make their bodies fit with new technologies and changing social ideas about being in and moving through public space. Along with a spectrum of ideas in popular periodicals, growing access to cycle-oriented sewing patterns helped women imagine and make new garments. Patricia Campbell Warner writes about how in 'November 1895 *The Delineator* [a monthly American women's magazine] published three pages of "Bicycle Garments" (all illustrated), offering readers some fifty Butterick patterns.'[46] This did not mean sewers were restricted to fixed or set styles. Sarah Gordon explains how patterns were 'designed to be interpreted in different ways, allowing readers to create their own definitions of what was appropriate and feminine.'[47] The lack of a singular accepted cycling garment meant even more freedom for makers. This created an atmosphere of individual creativity, whereby early cyclists shared ideas and patterns and discussed and debated different designs at events, in personal correspondence and various publications. As Lady Greville explains: 'I have read in cycling papers many descriptions of other women's bicycling costumes, but never yet have I discovered one which, for simplicity and appropriateness could complete with mine.'[48] Many columnists in popular periodicals also encouraged new cycle wear makers. Marguerite, of *Bicycling News*, was an informative source of knowledge:

> If any of my readers are going in for making a cycling costume at home, let me recommend to their notice 'Mrs Leach's Practical Dressmaker' for May, price two pence. They will find several very pretty styles illustrated therein, and, in addition, instructions for making up one style of dress which should prove very useful.[49]

The Lady Cyclist also regularly showcased patterns for home dressmakers. Madame Mode's Paper Patterns were in particular demand:

> During the first year of The Lady Cyclist's existence there has been a steady demand from its readers for the paper patterns of cycling costumes which are supplied by Madame Mode, the writer of the monthly fashion article, and we are pleased to announce that this feature will still be continued, the price if the complete cut-out and tacked-down patterns being still 1s. 6d. each, post free.[50]

Women also found inspiration in broader visual culture of the time. When women cyclists grabbed the headlines, their costumes often garnered as much attention as the record of their achievements. Not all of this attention was negative. Sixteen-year-old Tessa Reynolds hit the news in 1893 for a daring endurance cycling feat (she cycled from Brighton to London and back in a record time of 8 hours and 38 minutes for a distance of more than 100 miles) and also for her home-made costume that was 'closely approximating to that of a male person.'[51] It was featured in the 'Ladies Page' of *Bicycling News* under the title 'The Costume of the

Future'. Tessa's costume was a rational-dress-inspired coat and pair of bloomers, which she designed and made herself. 'I have received many applications for patterns of my suit from ladies ... I have not a pattern for it as I cut it out and made it entirely from my idea of what was wanted.'[52] It generated so much attention that G. Lacy Hillier, writing in *Bicycling News*, called her 'the stormy petrel heralding the storm of revolt against the petticoat.'[53]

Like Tessa, some women eschewed paper patterns and made up their own designs. Others re-configured garments at hand. The results were not always to everyone's taste. A writer from *Heart and Home* was shocked to witness a cyclist riding 'in knickerbockers of a violently checked pattern, made at home from a pattern provided by the old garments of their brothers.'[54] Aware that a badly designed costume could 'damage the cause', key supporters of rational dress, such as the Provisional Committee of the Rational Dress League, firmly directed women to seek the assistance of a reputable tailor.

> We advise, we entreat, all Leaguers to get only a tailor-made costume. The most expensive stuff, the daintiest idea is spoiled and useful if home-made or even dressmaker-made. The *cut* is everything. The suitable dress for London is a neat, plain and above all *well-cut*, tailor-made coat and knickers. The *cut* is everything.[55]

So passionate was their desire to prevent 'the wearing of slovenly or clumsy home-made costumes', that the League made arrangements with local tailors, such as R. Marcus of Alfred Place, Bedford Square who had agreed to make a cycling costume that addressed the League's brief for three pounds and ten shillings which could be paid in full or monthly installments. Marguerite also advised her readers to seek professional help, but she recommended getting a dressmaker who was also a cyclist:

> I would advise such ladies to get their riding costumes made by a practical cyclist where this is possible. A little riding experience makes a wonderful difference, and, if you know of a dressmaker who rides, by all means let her make your dress.[56]

A similar sentiment was voiced by the editor of *The Lady Cyclist*, who warned the novice against favouring fashionability over practicality. 'They are so ready to accept the dressmaker's or tailor's word that

Figure 3.4 Portrait of Tessie Reynolds in her radical cycling outfit, *Bicycling News*, 1893

the skirt is suitable for cycling, and only ask for something fashionable, when, of course, they get a design which looks charming on the tailor's lay figure, and which is a regular death-trap on the bicycle.'[57]

Regardless of whether women made their new cycle wear themselves or commissioned someone else, they still had to know what they wanted. Many saw this as an opportunity to explore and experiment and some entrepreneurial individuals, many of whom were women, also patented their ideas. It is through their patenting activities that we have a chance to better understand their motivations and imaginings of a different cycling world.

4

The 1890s Patenting Boom and the Cycle Craze

The craze for bicycling has made a complete revolution in the needs of dress, and there are almost as many inventions for this special amusement as days in the year. Every week at least a patent is either taken out or applied for touching on bicycle clothes. Happily there is a variety of opinion as to the requirements of this particular amusement.

The Queen, The Lady's Newspaper, 1895[1]

The Victorians are renowned for their engineering spirit and inventive legacy. The late nineteenth century marked a prolific period of new inventions. In part this was due to changes to the patenting process which opened the system to new inventors, who previously had little chance of gaining entry into this exclusive world. A patent turned 'an idea into a form of property'.[2] An idea became a legal entity that was recognisable, defensible, valuable, consumable, and, of course, sellable. News of patents in the 1890s seemed to be everywhere – in the papers and advertisements, the source of much gossip, on the streets and in the theatre and splashed across a huge range of new products. In many ways it was the equivalent of the dotcom boom a century later. Patent fever captured the collective social imagination, seducing potential inventors with fantasies of previously unobtainable economic transformation and social mobility. On a broader scale, it was fuelled in part by England's determination not only to be part of the 'great race' to modernity, but also to lead it into the future. Along with the colonial project, transforming simple ideas into great inventions was a compelling way of doing just that. A close look at Victorian legal systems and social context provides insights into why individuals sought to patent their designs and why some ideas were considered more valuable than others.

The race to patent ideas at the turn of the last century is full of drama and excitement, late nights in the patent reading room, frenzied dashes to the patent office, about luck and connections, travels across oceans to sell ideas, demonstrations to attentive crowds, newspaper journalists waxing lyrical amidst flashing bulbs of photographers and, of course, surviving the inevitable descent when the spotlight moves to the next big thing. Feminist Technoscientist Ruth Schwartz-Cowan writes about the lives and motivations of infamous American entrepreneurs such as Samuel Morse (telegraph), Thomas Edison (electric light, photography) and Alexander Graham Bell (telephone). Successful inventors like these became celebrities, their names indelibly forged in national memory. 'Newspapers quoted their opinions; popular magazines recounted their exploits; huge crowds turned out to hear them lecture; artists clamored for the right to paint their portraits.'[3]

Britain in the mid-1890s witnessed a particular form of patenting fever. Many of the ads in popular periodicals and newspapers proudly featured the 'Patent' behind new designs. A patent was a key selling proposition. It meant brand new and cutting-edge. The number of patent applications rose by 20 per cent in one year alone. In 1897, the *Annual Patent Report*, which provided regular updates for Parliament, confirmed what was well known in the media of the time – the increase was attributed to the cycling craze sweeping the nation.[4] This was not just focused on the velocipede itself but spilled out into the many components that surrounded it. There was something approachable about this technology that tempted all manner of inventors to try their luck. It was not hard to see why. The late nineteenth century had brought forth many radical changes with the invention of toilet paper (British Perforated Paper Company, 1880), the automobile (Gottlieb Daimler, 1886), escalator (Jesse W. Reno, 1891), diesel engine (Rudolf Diesel, 1892), zip (Whitcomb Judson, 1893) and the movie camera (Lumière brothers, 1895), to name a few inventions that have indelibly shaped the world we live in today. More motivating perhaps to small potential inventors were stories of successful minor inventions. A column in *The Church Weekly*, a London newspaper, unambiguously titled 'Inventions Which Have Made People Rich' starts with the sentence: 'Most of the big fortunes earned through patents have been gained by small things, such as would not have considered important by the casual observer.' Examples included a child's toy which yielded £16,000 in its first year. A glass lemon squeezer generated £10,000 and there was apparently £100,000 to be made from a wooden shoe peg.[5] Profitable ideas seemed to be everywhere.

The opportunity to radically alter your social circumstances would have been particularly appealing for women. They were much less likely to have had an uninterrupted period of schooling or training. Even those with this privilege still suffered discrimination associated with being the 'weaker sex', which limited their ability to do much with their education. In the early part of the century, middle- and upper-class women's lives were firmly attached to the home, not the workforce (this was very different for lower-class women who worked throughout their lives). Born into or marrying money via a 'good match' remained the primary path to a good life for many. Neither offered women much in the way of precious financial independence.

Things were changing in the late nineteenth century, opening up for women the possibility of considering alternate livelihoods. Lee Holcombe's study of *Victorian Ladies at Work* illustrates the political moves made by emancipists to advocate for women's opportunities to engage more actively in the public sphere. Some took the position that women were not rejecting conventional beliefs outright but seeking to adapt them to the benefits of everyone: 'Besides creating better daughters, wives and mothers, the feminists argued that improved education and opportunities for women to work outside the home would benefit society at large.'[6] Popular periodicals also began to showcase the growing range of opportunities to expand the potential of modern feminine lives, and convince those around them to not resist these changes. *The Queen* was one such publication that featured a regular column, 'Women's Employment', listing recent appointments, answering readers' queries and publishing first-hand accounts of what it was like to do different kinds of jobs as well as the type of training required to gain such positions.

The practical necessity for middle-class women to work was also growing. Divorce rates were on the rise, so women from a range of classes were looking to find new forms of support outside their marriage.[7] This led one newspaper to declare: 'Many a woman who would have clung to a worthless husband from dread of starvation did she leave him, now reasons that she can make a better living by freeing herself from a tiresome encumbrance and going to work in a shop or factory.'[8] Working in a shop or factory, however, was not as liberating as it might have first seemed, even in comparison to dreary domestic life. Conditions were often poor and pay was not much better, especially for women who earned in some cases half of men's wages for comparable work.[9] More opportunity was available for middle-class and educated women to go into teaching, nursing, dress-making and other similar jobs 'considered to be women's work.'[10]

Regardless of class, the thought of becoming an inventor must have held almost unimaginable promise of a independent life and good income. It offered recognition and even prestige but most of all perhaps, a taste of freedom. But it was initially complex and costly to patent an idea. For women it was even more difficult, owing to their lack of legal rights, education and networks.

Patent Reform and the Cycling Revolution

There are a number of factors that opened patenting up to a broader body of inventors, including women. One of these was reform to the patenting process that took place in the 1880s. The 'Patents for Invention Bill' was read and debated for the second time on 15 June 1881 in Parliament, and reported in *The Times*.[11] The main argument related to whether the current patenting system was beneficial to the public, or only for some inventors. Did the fees and length of patents 'discourage invention'? Was the process a barrier to new ideas? Comparisons were made to patenting laws in other places, particularly the United States, which was viewed as 'infinitely superior.' There, the proportion of patents was one in 3,000 of the population. In Britain it was one in 12,000. The idea of being left behind in a global context was powerfully motivating. During the preceding 28 years, British patent applications had increased, but very slowly. There were 1,211 applications in 1852 and by 1880 this had risen to 6,000. These application rates compared unfavourably to other countries, including the United States, which were close to three times higher. There, small patents were apparently achieving disproportionate rates of success and helping the country forge a reputation of being especially inventive in 'the great race'. This was deeply worrying to British politicians:

> Mr D Grant observed that the foundation of the great success of the American nation was the use by them of labour-saving apparatus, which was the outcome of a number of small inventions. If we were to win in the great race we were entering upon we must remove the present tax upon invention.

These kinds of stories from abroad were starting to change the view of the patent process at home. Perhaps Britain should be more open to a diverse range of inventors and inventions, wondered *The Times* journalist who was covering the political debate:

> A few years ago a general feeling appeared to have prevailed against patents, and many of the most influential statesmen in all countries seemed to be opposed to them upon principle. That feeling, however, had undergone a change of late years, and now the view entertained was that patents should be encouraged rather than discouraged. It was, however, generally admitted that our Patent Law was in such an unsatisfactory state that a thorough reform of it had become necessary.[12]

The differences between the British and American systems were debated at length. Many parliamentary interlocutors stressed the need to support and encourage smaller inventions that could have a big impact: 'The American system of patents was infinitely superior to that of this country, inasmuch as it enabled patents to be taken out cheaply for small improvements'. However, this required a simpler, less cumbersome legal process.

Kara Swanson's research into *The Emergence of the Professional Patent Practitioner* focuses on the role and 'meteoric rise' of practical experts paid to assist inventors prepare and submit patents. She explains how the 'U.S patent system was frequently lauded as cheaper, simpler and more effective at issuing valid and valuable patents than the British system'.[13] The process cost only £7 in the United States. In comparison, under the Act of 1852, the British patenting process involved seven applications and four payments. The first was £25, with an additional £50 required before the third year and £100 before the seventh. This presented a clear barrier to entry for inventors, lacking personal wealth or a supportive patron. As noted in *The Times*, this caused a significant drop off 'either because the inventor was unable to pay the fees, or because in the interval he had discovered that the patent was of no value'.[14] Only 11 per cent of all patents proceeded past the seventh year. The length of patents was also debated. The British patent privilege was far shorter than in the United States, at 14 and 17 years, respectively. Even the quality of the British patent office wasn't safe from criticism: 'The Patent Office at Washington was one of the finest buildings in the city, in striking contrast to the miserable structure in Chancery-lane.' Ultimately, a more thorough review of local and international systems, such as that of the American process, was deemed essential.

Women were one group, amongst many, disproportionally disadvantaged by the traditional patenting system. An 1894 article in the *New Scientist* titled 'The Innovative Woman' queried why so few women had patents to their names and identified several fundamental barriers: 'Invention usually requires money, materials and the opportunity to share ideas', it argued, and 'historically, few women have been financially independent, and most have been excluded from sources of education and intellectual stimulation'.[15] It was not only the cost and a supportive community that inhibited their ability to patent ideas, they were also greatly constrained by their legal status. The Married Women's Property Act came into force in 1870, but it wasn't until 1882 that it included women's right to legally own property in their own names. Prior to this, they were considered dependent on their fathers, brothers and husbands. Presumably, women were responsible for many innovations up until this time that were either not recognised or claimed by men in their lives.

Shortly after the parliamentary debate, the Patent, Designs and Trademarks Act was passed in 1883 with the aim of 'simplifying the methods of obtaining, amending, extending and revoking patents'.[16] This noticeably opened up the process to different types of inventors and inventions. The cost of submitting an application plummeted from £25 to £4 and involved only two applications; a provisional protection (£1) and a complete specification (£3). There were further fees of £50 before the end of four years and more again, £100, after seven years.[17] However, these changes meant that an inventor could claim an idea for an initial period for only £4, and then pay to extend if needed.

This new initial low entry fee proved popular. The first day of 1884 saw the largest volume of patent applications in any day, at 266, and a total of 2,499 were submitted in January. The usual monthly average was 500. The reform was an undeniable success. Applications for the year reached 17,110, a whopping increase on the previous year. The Comptroller General writing in the 1884 Patent Report clearly thought this answered the question posed in the parliamentary debate, though he presented it in an entirely understated manner: 'The new Act may be said to have worked well in the interests of inventors' as evidenced by an 180% increase in patent applications.'[18]

Patent applications continued to grow steadily through the following decade: 18,051 were lodged in 1887, 21,307 in 1890 and 25,120 in 1893. While women had been inventing throughout this time, in spite of the barriers, they suddenly became visible to the establishment in the 1890s. Their patenting activity started to become statistically relevant. The volume and type of patents were remarked upon for the first time in the 1894 Annual Patent Report: 'Of the 25,386 applications received in the year 1894, 501 or two per cent., were made by women, about 100 being inventions connected with articles of dress.'[19] Women's patents increased again in 1895, and again it was noted: 'Five hundred and ninety-one, or 2.3 percent., of the total number of applications, were made by women during the year; about 184 being for inventions connected with articles of dress.'[20] By far the most impressive jumps in patenting overall, and also in terms of women's involvement, were still to come.

Aside from the Patent Reform Act, and for women, the Married Women's Property Act, another catalysing factor underpinning the growth in patents was the frenzy generated around the craze of cycling, which was far-reaching through all levels of society. The year 1896 saw applications rise 20.5 per cent to 30,194. To put it in context, this was a similar leap to the impact of the Patent Reform Act in 1883. Again, this leap was recorded in the Annual Patent Report: 'The principal cause of the rise in patent applications is to be found no doubt in the development of the cycle industry, to which more than 5000 of the inventions reference.'[21] The rate of women's patenting continued to be noted: 'Six hundred and ninety-one, or 2.3 per cent., total number of applications were made by women during the year; about 153 being for inventions connected with articles of dress.'[22] However, what the report fails to mention is if these 'articles of dress' were cycling-related. Patents for cycle wear were not categorised under this rapidly expanding classification of *Velocipedes*, but rather under *Wearing Apparel*. This is an important point, because the former generated far more media coverage and speculator attention than the latter. It partially accounts for why women's inventions were less recognised and championed as being part of this massive industrial revolution and arguably continue to remain virtually unknown today.

The year 1897 was also critical in patenting history. Applications rose a further 2.5 per cent to 30,958, with another 6,000 new cycling-related inventions.[23] This year the Annual Comptroller Report states that while the increase was smaller, the quality was higher as there were more patents with complete specifications. Fears the parliamentary committee held about the increase in 'frivolous patents' did not materialise. As noted in the report: 'When the Patents Act of 1883 first came into force, and the initial fees were reduced from £25 to £4, it was thought probable that the average value of the Patents granted would be diminished in corresponding degree, as trivial inventions, upon which it had not been worth while to pay high fees, would in future be made the subject of Patents.' Yet, this was not the case. Numbers for patents continuing for their full term were 'considerably larger than before.'[24] This report also starts to note more detail about women's patents, not only for dress but also for cycling. 'Women inventors contributed 702, or nearly 2.3 per cent. of the total number of applications, about 148 being for inventions connected with articles of dress, and 106 for inventions related to cycling.'[25] Again, it does not make it clear if the articles of dress were related to cycling.

The next year, 1898, saw overall patents drop 10.7 per cent to 27,659.[26] Although 6,000 of these were still cycle-related, the decline in the popularity of cycling was blamed: 'The rapid growth in the number of Specifications which took place in 1896 and 1897 was ascribed to the activity of the cycling industry, and there is little doubt that the industry has been principally responsible for the present decline in numbers.'[27] Despite this drop, women continued to submit patents at nearly the same rate. 'Women inventors contributed 683 or more than 2.4 per cent of the total number of applications, about 148 being for inventions connected with articles of dress, and 79 for inventions related to cycling.'[28] Again, there is ambiguity in the categories.

The year 1898 also saw an increase in patents being lodged in Britain from Austria, Denmark, Germany, Italy, Norway, Russia and Sweden.[29] A patent agent reported the change and although disappointed by their home country statistics, was nonetheless pleased to be living in this age of ideas. Messrs. Stanley, Popplewell and Co. of 61 Chancery Lane write: 'It may then be concluded that, notwithstanding the decrease in the number of English applications, the world of invention is as active as ever.'[30] The agent also notes with surprise women's continued engagement with the patent process and the diversity of their inventive capacity:

> Women inventors contribute some hundreds of patent applications yearly, the proportion continuing with strange regularity at 2.3 per cent. of the whole. The subjects include dress, cycles, and even mechanical and engineering devices.[31]

Women Inventors Fight to be Recognised

By the late 1890s, the idea of women inventors began to slowly capture the Victorian imagination. Although they were still lacking political and economic rights there was significant evidence of female ingenuity in patents. Newspapers and periodicals were starting to report on women's inventions in the context of happenings in the larger patent world. Informal channels of communication also circulated news of ideas.

The first issue of *The Rational Dress Gazette* published in June 1898, highlighted the importance of inventors in one of its key objectives: 'To encourage inventors, makers and manufacturers of improvements in dress and dress materials, and to provide by exhibitions and other means a channel for communication between said makers and Rational Dress wearers.'[32] The *Gazette* lived up to these aims by regularly encouraging the sharing of ideas and patterns, names of new patent holders and reviews of their garments.

Advice also flowed across continents between members of emancipist groups. *The Dawn: A Journal for Australian Women* was a feminist periodical published in Sydney from 1888 to 1905, set up by Louisa Lawson, a writer, poet and renowned feminist (and mother of famous poet Henry Lawson). In an 'Answers to Correspondents' section published in 1896, an L/A from Bega in south eastern Australia offered advice to a previous letter writer who had asked about the American patenting process: 'It costs twenty pounds to obtain a patent for any article for the United States of America if done through a patent agent.'[33]

Women's inventions and inventiveness became a popular subject of lectures and opinion pieces in periodicals and newspapers. A column in *The Church Weekly* titled 'What Women Have Done' reported in 1899 on a recent lecture by an American writer, Mrs Bowles, called 'Women as Inventors', that referenced 12 years of research into women's textile work and patenting history.[34] Mrs Bowles told the listening crowd that the first patent by a woman, in 1808, was by Mary Kees, for weaving straw with silk or thread. There were 15 patentees over the next 25 years. 'Among these inventions were a globe for teaching geography, a baby jumper, a fountain pen, a deep-sea telescope and the first cooking stove.' The major barrier she saw was women's lack of education, but this had changed over time 'when more privileges were accorded women' and the volume of patents increased correspondingly.

The *Church Weekly* article reports on how some women held more than one patent and the diversity of their inventions crossed a spectrum of domains. Mrs Harriet Strong, for instance, patented an improved corset as well as designs relating to reservoirs and irrigation to address water storage and flood issues. The article explains how these ideas emerged from personal experience and practicality; the former from her constant back pain and the later as the result of her husband's death. 'Catapulted into the economic area by her husband's suicide in 1883, she moved with her four daughters to a debt-ridden ranch near Whittier, C.A, and set out to make it pay.' These examples clearly evidenced women's critical and creative abilities.

Women were inventing and laying claim to their ideas, despite the legal and social challenges they faced. Further to these newspaper references, patent records provide clear and irrefutable record of this. As Ruth Schwartz-Cowan has argued: 'If there were no such thing as a patent, we would not know very much about inventors.'[35] Patents are also a particularly good record of women's inventions, as Zorina Khan writes: 'Patent records present a valuable perspective on female inventive activity and market participation in an era when marriage meant the virtual "invisibility" of married women in terms of objective data.'[36] Moreover, the chronological nature of patents means we can map women's patenting activity against other social and political happenings, such as the Patent Reform Act and the Married Women's Property Act, to consider the impact different rulings had on opening up and closing down women's inventiveness. They provide a unique channel for women's creativity and critical design faculties, and enable us to hear directly from them at a time when they had few formal platforms for expertise and knowledge sharing. Moreover, they provide a solid record even if the inventions themselves were not actually physically made then or are not available now.

Yet, despite the evidence, women inventors, and their inventions, struggled for legitimacy. The *Church Weekly* article on Mrs Bowles's lecture concludes with an illustrative anecdote.

She was out driving with an old farmer and he said to her testily:

'You women may talk of your rights, but why don't you invent something?' to which Mrs Bowles quietly replied: 'Your horse's feed bag and the shade over his head were both of them invented by a woman.'

'You don't say so!' was the amazed rejoinder.

Women were more often seen as 'imitators', supporters or followers, than as creators of new ideas. As the editors of *The Queen* argued, at the height of patenting fever in 1896, recognition seemed ever elusive: 'It is one of those numerous generalisations about feminine capacity which are accepted without much consideration – that women are not inventors. Imitators, both clever and ingenious, they are freely allowed to be, creators never.'[37] A similar maxim was discussed and disputed in *The Dawn*, four years earlier. Again, it was accompanied by clear evidence to the contrary. Women *were* inventors.

> We have often been told that women possess no inventive faculties, so we were glad to note an invention by a woman, the usefulness of which falls within her special sphere, inasmuch as it tends to increase materially the comfort of the helpless and suffering invalid, as the doctors who have used it are ready to testify. It consists of an ingenious compound of an ordinary hair-mattrass with a large air cushion, which can be emptied or inflated without disturbing the patient.[38]

There is also further evidence that patent agents and solicitors who assisted inventors with processing their applications through the legal system recognised women's inventive potential and were supportive of them entering this domain. In a column titled 'Female Inventive Talent' in the *Scientific American* in 1870, a writer argues the case for inventive women:

> We have frequently been called upon to prepare applications for female inventors, and to correspond with them in relation to various inventions; and we can say to those who are unbelievers in regard to the power of women to achieve, as a class, anything higher than a pound-cake or a piece of embroidery, that the inventions made by women, and for which they solicit patents through our agency, are generally found to be in their practical character, and in their adaptation and selection of means to effect a definite purpose, fully equal to the same number of inventions selected at random from among those made by men.[39]

So, why were they not recognised then and why, given how few female inventors are well-known today, does this perception still exist? One answer to this lies in what constituted a real, legitimate invention. This was far from fixed or stable, even if the inventor had successfully patented their idea. Further instability for women lay in the categorisation of their inventions. Zorina Khan notes how organisers of the 1893 Women's Pavilion for the World's Columbian Exposition asked 'to make no note of the inventions of women unless it [was] something quite distinguished and brilliant'. The reason for this was fear that anything less would be harshly criticised and easily mocked. 'We must not call attention to anything that would cause us to lose ground.'[40] As a result, they regulated what 'counted' as 'proper' invention. By this, they meant nothing to do with traditional women's arts and crafts, household or domestic interests. As Zorina Khan writes:

> Even the Women's Bureau Bulletin documenting women's inventions from 1905 to 1921 opined, '[I]f the steady increase in the numbers of patents granted women is accounted for merely by the increase in the number of patented hairpins, hair curlers and such trifles in feminine equipment, it is without larger significance either to civilisation or as an indication of women's inventive abilities'.[41]

Inventions often emerged from an individual's personal experience with a task or activity. Given women were predominantly limited in terms of their access and engagement with public space in the nineteenth century, it was often the case that their inventions coalesced around the home, clothing and family life. These inventions were not highlighted as valuable within either society at large or, it seems, by some women's rights activists concerned about limiting women's potential, and yet doing just that in their dismissive attitudes. This had the effect of creating yet another barrier to entry for women, many of whom were coming up with a plethora of inventive solutions for the problems they identified in the spaces they inhabited. As Zorina Khan writes: 'By denigrating household work and the inventions of household articles, the women's movement likely contributed to the notion that women were not technologically adept.'[42]

Yet, conversely, some women were also criticised for venturing outside their 'special sphere'. Frustrated by these limitations on women inventors, a writer in *The Queen* notes with sarcasm: 'Women have sometimes shown a shocking tendency to allow their inventive faculty to wander quite out of the feminine sphere.' They go on to list patents by women 'for caulking ships, boats, and other vessels' and 'improved hauling-out slipway.' They ask why women's inventiveness in these more masculine fields was called into question when men had been laying claim to inventions in the feminine sphere without it causing social upset. 'Yet, from the time when printed records were first kept (1617) up to 1852, not a single woman acquired a patent for "sewing, embroidery, and tambouring", neither for spinning – numerous patents of which have been taken out by men – nor for ornamenting of anytime.'[43] This argument is further problematised by the fact that men had most likely, due to various systemic and social barriers, been claiming responsibility for inventions by women for a long period of time.

This matters, because even a brief glance at the histories of technological development would have readers believe that all spirited historic inventors were male. Yet, this was clearly not the case. Feminist technology scholars like Judy Wajcman have written prolifically about the lack of women in technological histories: '[T]heir absence is as telling as the presence of some other actors, and even a condition of that presence.'[44] Women's ability to manage home and family life, and perform critical supporting roles, enabled men in many instances to forge technological inventions, yet they rarely gain a mention in the annuls of history. Ruth Schwartz Cowan has also written compellingly about how 'the absence of a female perspective in the available histories of technology was a function of the historians who write them and not of the historical reality.'[45] A ramification of this is how we know more about the bike than the baby carriage, more about harvesting technology than the playpen and more about power looms than the baby bottle.[46]

Although there is evidence that patent agents recognised and supported women inventors, the process of patenting may still have been daunting to some women, due to the absence of female staff working in patent offices. Looking through British Annual Patent reports reveals that, with the exception of a charwoman (a low-paid cleaner), all of the employees of the patenting office throughout the late nineteenth century were male. The intimidating experience of bringing an idea to be patented into an all-male office, especially given the real possibility of scorn, cannot be discounted. Zorina Khan notes how this was recognised and addressed in the United States when 'in the 1870s, the Patent Office hired its first female patent examiner, possibly encouraging women to submit inventions that they might have feared would be viewed with less sympathy by male examiners.'[47]

For those fortunate enough to have their ideas successfully patented, how their inventions were categorised remains relevant to their long-term legacy (or lack thereof). Women's patenting during the boom years of 1896 and 1897 was predominantly around wearing apparel *for* cycling. In addition to the bicycle itself, the 1898 Patent Report made special mention of the associated inventive categories it viewed as having substantial impact on the market – '*wheels* (including pneumatic tyres), *bearings, chains* and *air and gases, compressing* (tyre inflators)'.[48] It does not mention clothing. The patent category of *Wearing-Apparel* also shows a marked increased in the period 1895 to 1896, comparable to the categories listed above. But clothing inventions, even while predominantly around cycling, were not associated with *Velocipedes*. In the abridgements, an annual report of patent abstracts for easy overview of each year's inventions, *Wearing Apparel* includes a category for *Cyclists' Wearing-Apparel*, but this is even less clear. Here patents are distributed again into multiple sub-categories: 'Capes; Cloaks &c.; Corsets; Dresses &c.; Dress-improvers; Gloves; Leggings &c.; Trousers &c.; Under-vests &c.; Waterproof garments'.[49] Readers have to piece together fragmented data to find cycle wear within these disparate sections.

Looking closely at classification systems is important because at the time they shaped how patent news was distributed, circulated and, ultimately, valued. Excerpts from the Annual Comptroller Patent Reports along with lists of recent patents appeared in broadsheets and popular weekly periodicals. Sometimes, these listings were paid for by local agents who used them to promote their skills and services. An example reads: 'This list is specially compiled for "Bicycling News" by Messrs. Rayner and Co., registered Patent Agents, 37 Chancery Lane., London, WC.'[50] These regularly published lists featured recent cycling inventions such as improvements in brakes, pneumatic tyres, gearing, cycle frames, toe clips, saddles and cyclometers. They did not include patents for cycle wear. This meant that bicycle inventions gained much more attention than clothing in the 1890s. Had women known that other women were inventing things, it might have catalysed more invention at the time, and perhaps we would know more about them now. Writing about the impact of the female patent agent in the American system, Zorina Khan agrees: 'Women may have been prompted to invent and patent by the example of other successful female patentees.'[51]

Science and Technology Studies scholars have written about the critical importance of classification systems in everyday life and how they are often undervalued.[52] In *Sorting Things Out: Classification and its Consequences* Susan Leigh Star and Geof Bowker question why this is the case: 'Remarkably for such a central part of our lives, we stand for the most part in formal ignorance of the social and moral order created by these invisible, potent entities.'[53] Classification is both a mundane daily activity and dominant infrastructure that shapes privilege and power. You only have to consider the power of maps for their potential to simultaneously convey and conceal knowledge. They are political objects, drawn in many cases by the victor, with an explicit purpose in mind. It is easy to overlook and accept knowledge systems, especially when they look official and definite.

Patent archives are particularly productive and problematic places for (re)thinking about categorised knowledge. Here, knowledge appears firm and settled in taxonomic order. Yet, a recent 'archival turn' in the social sciences has, as Kate Eichhorn explains, 'made it commonplace to understand the archive as something that is by no means bound by its traditional definition as a repository for documents.'[54] Her research into and about the archive explores how different views of the same materials can render new entry points and insights. 'Rather than a destination for knowledges already produced or a place to recover histories and ideas placed under erasure, the making of archives is frequently where knowledge production begins.'[55] Similarly, Ann Stoler's research into colonial archives questions not only classification systems but also the larger epistemological systems at play. She calls for a 'move from the archive-as-source to the archive-as-subject'.[56] Reflecting on this work invites us to think about archives as already deeply politicised entities shaped by social, cultural and gendered beliefs and assumptions.

In the case of Victorian cycle wear patents, records were fragmented and categorised in and across different places. The most obvious stories and characters we learn about are not the only ones that should demand our attention. There are many others. If we consider the taxonomy of the patent archive as a starting point for investigation, then we can ask: why are stories told in these ways? What and who might be missing? How else might we piece the fragments together? Might it be possible to summon into the present a different account of Victorian inventiveness from the past?

5

*Extra*ordinary Cycle Wear Patents

In the old days of tricycles, and when they were not at all the fashion, cycling dress was not the fine art it is now. Our only idea was to look neat, and be clad in weather-proof garments. Sailor hats in summer, and felt hats in autumn and winter, were the sum of elegance required, and fashions – Paris, rational and otherwise – did not require to be studied. We have changed all that now.

Miss F.J. Erskine
Lady Cycling: What to Wear and How to Ride, 1897[1]

What new kinds of cycle wear were being invented? Who was patenting it? Where was it being worn? The 1890s brought with it remarkable changes to clothing designed for the purposes of cycling. Victorians, and particularly women, were engaged in experimenting with a range of new kinds of clothing, to enable women to embrace more active lives, that were not necessarily divided into *rational* or *irrational* dress. Instead, many designs occupied a dynamic space in-between; that were both and neither. Some of these radical new costumes were deliberately designed to operate as cycle wear and yet also evade a definitive label. These kinds of patents open up discussions around women's cycling dress beyond that of the conventional binaries; of feminine and masculine identities, of mobile and immobile bodies, of those who wore rational dress and those who resisted, and of familiar dimensions of success and failure.

As Miss F. J. Erskine exclaims, cycle wear had changed dramatically, to the point where it was now a 'fine art'. A broad spectrum of garments addressed a plethora of cycling needs and wants, which as Miss Erskine explains, spanned from 'Paris' to 'rational' to 'otherwise'. Enthusiasm for cycling, combined with awareness of social norms and changing political climate, motivated many to customise costumes to fit a new landscape of possibilities and necessities. The fact they were patented usefully allows us, over a century later, to gain insight into what inventors saw as problems and how they sought to fix them. In their own way, each of these designs set out to enable women to forge new mobile identities. As such, patenting can be seen a political act, not only in creating a place for women's cycling bodies in public space but also in attempting to carve out new technical, social and gendered futures.

Themes in Patents

Eighty-six British patents for new or improvements to women's skirts for the purposes of cycling were registered between 1890 and 1900.[2] Many more patents were registered for a broader range of cycle-oriented clothing, such as footwear, blouses, coats and corsets, but I have focused on skirts and bloomers/knickerbockers worn underneath or in place of skirts. While this may not sound like a vast volume, these inventions marked a radical shift in women's clothing. Like others, Diana Crane has argued that change happens not in the centre but rather on the marginalised edges. 'Had a nineteenth-century social scientist set out to predict how women would dress at the beginning of the twenty-first century, it would only have been by considering the clothing of the most marginal women in Europe and America that an accurate assessment would have been obtained.'[3]

Most of the inventors of this period came from Britain, but there were also patents from Germany, Canada, Chicago, Minnesota, New York and as far away as Melbourne in Australia and 'the Colony of New Zealand'. Women submitted close to half. The men all identify with a trade; such as Tailor, Milliner, Cutter, Clothier and Costumier as well as Engineer, Solicitor and Commission Agent. Most of the women remain less vocationally identified, broadly listed as Spinsters, Gentlewomen, or wives of men who have jobs. However it is notable that there are three Dressmakers, a Milliner, Costumier, Composing Pianist, Dispenser and Registrar at a Hospital for Women, Governess and Lecturer on Hygiene and Physiology to the National Heath Society. The patents are diverse in terms of creative responses to the 'dress problem'. Yet, they shared a central concern – to design a garment that operated as safe and comfortable cycle wear while still providing the 'appearance of an ordinary skirt'. This was not a minor challenge. Ultimately they were attempting to produce something *extra*ordinary with the ordinary.

> This invention relates to a new or improved cycling skirt for ladies' wear which though divided and possessing all the advantages of the divided skirt has *the appearance of an ordinary walking skirt.*[4]

> This invention relates to improvements in cycling skirts and has for its object to construct these in such a manner as to allow the rider the full use of her limbs without any of the leg exposed and at the same time to *have the appearance of an ordinary walking skirt* when the rider is not on her machine.[5]

> This invention relates to improvements in connection with ladies' skirts and has for its object to provide an arrangement which can be *easily altered from an ordinary skirt* into a divided skirt and *vice versa.*[6]

These cycle wear patents can be mapped across five different categories (and linked to the strategies in the previous chapter). In addition to examples of patents in each, I also point out commercial garments that appear similar but were produced under different names. It was common practice to replace the inventor's name with a new brand, which sometimes makes it difficult to keep track of original inventions. Nevertheless, these categories reveal some of the larger issues and anxieties that shaped inventors' creative endeavours.

Device to Attach, Stiffen or Secure Skirt

This design solution directly addresses the movement of the skirt while cycling. Inventors set out to prevent the skirt from flapping around in the wind and gathering up over the knees with the pedalling movements of the legs. These patents include straps, buckles and metal belts that weighed a dress down or fixed it to the rider's body. The 'Fixit Dress Holder' is an example of a patented commercial garment that was promoted as an alternative to rational dress. An advertisement in *The Lady Cyclist* proclaimed: 'By using this Dress Holder Ladies may cycle in comfort in ordinary walking dress and their skirts will be held down in the strongest wind.' It must have worked to some degree, as it had glowing testimonials from popular magazines *The Gentlewoman* and *The Ladies Gazette*. The Fixit comprised a silk or cotton elastic loop which fixed the skirt to the ankle and apparently kept the material from blowing up. Designs like this stemmed from the well-known practice of adding weights to the hem of jackets and skirts. A review also appeared in *The Lady Cyclist*'s 'Letters of a Lady Cyclist':

Figure 5.1 The Fixit Dress Holder, advert in *The Lady Cyclist,* 1896

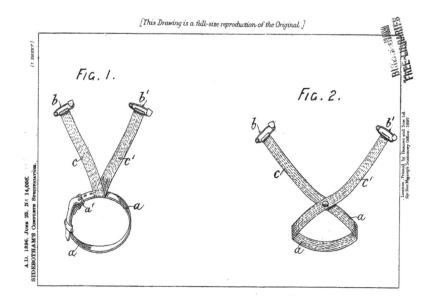

Figure 5.2 Lily Sidebotham's patent for 'An improved Appliance for Keeping Dress Skirts in Position While Cycling'

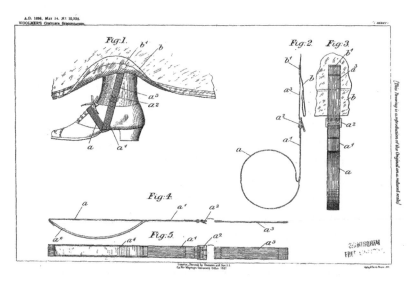

Figure 5.3 Emily Woolmer's patent for an 'Improved Skirt Holder for Lady Cyclists'

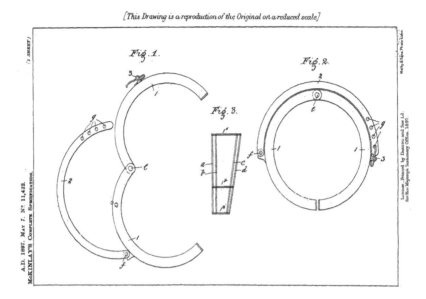

Figure 5.4 Alexander McKinlay's patent for 'Improvements in Ladies' Cycling Dress Protectors'

> A few weeks ago I sent for a pair of 'Fixit' dressholders. They are a most delightful invention, and have proved to me finally that rational dress is quite unnecessary. You must certainly have a pair, and your dress won't blow up in the least.[7]

The Fixit could well have come from a number of patents submitted around this time (and inspired those that came after). Lily Sidebotham, wife of George Henry Sidebotham, a Draper of Newport in the County of Salop, patented 'An improved Appliance for Keeping Dress Skirts in Position While Cycling' and declared the benefits of her design were to prevent the skirt from 'rising or otherwise becoming displaced in an uncomfortable or unsightly way'.[8] A similar patent, with clear illustration of how the skirt was strapped to the cyclist, is provided by Emily Christabel Woolmer, a Spinster, of The Vicarage, Sidcup.[9] Other patents included the use of leather, chains or elastic devices to attach the dress to the cyclist. Some were surprisingly complex and heavy. In his patent for 'Improvements in Ladies' Cycling Dress Protectors', Alexander McKinlay of Manchester proposed a system of four girdles with steel supports on a hinge mechanism.[10] Once closed, the stiffness of the girdles kept the dress in place, and given the weight, probably kept the woman away from her bicycle as well.

Tailor Skirt to Fit Bike

While the above patents generally *added* elements to skirts, this design solution sought to *remove* excess weight and material. Inventors in this category focused on reducing the dangers of a flyaway skirt by tailoring it to fit the bicycle. This included lessening the bulk of the garment, cutting it to fit over the back wheel or adding hidden gussets to enable the movement of the legs without causing the raising of the skirt over the knees. When away from the bike, these revisions remained undetectable.

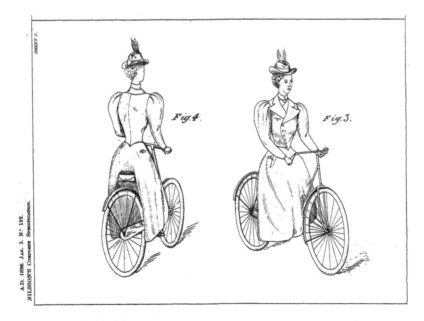

Figure 5.5 Peter Nilsson's patent for 'A new or Improved Cycling Habit for Ladies Wear'

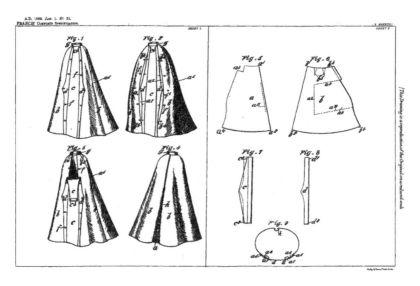

Figure 5.6 Susan Emily Francis's patent for 'An Improved Cycling Skirt' with adaptable gusset action

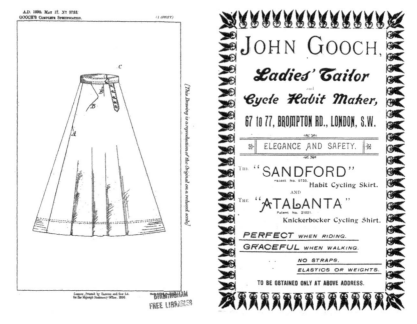

Figure 5.7 John Gooch's patent for 'Improvements in Ladies' Skirts or Dresses for Cycling' and an advert for his patented 'Sandford' Habit Cycling Skirt Patent, *The Lady Cyclist*, 1896

Figure 5.8 A Parisian divided skirt, *The Cycling World Illustrated*, 1896

In 1896, Peter Nilsson, a London Ladies' Tailor, patented 'A new or Improved Cycling Habit for Ladies' Wear.'[11] He explains how his design 'allows for sufficient amount of fullness below the hips to allow of requisite freedom in pedaling, but it is not to be so full as to leave a surplusage that in gusty weather can be blown about.' Again, like the Fixit Skirt Holder, Peter offers his design as a direct alternative to the controversy surrounding rational dress. 'This invention relates to a new or improved cycling costume or habit which whilst affording that freedom to the wearer which is necessary, will not present the objections which are found to the use of the so-called rational dress.'

In the same year, John Gooch, a London Outfitter, patented 'Improvements in Ladies' Skirts or Dresses for Cycling.'[12] His invention acknowledges the 'discomfort' and 'considerable danger' of conventional fashion that 'arises from the fact that the ordinary skirts are too full and loose.' So, he set about to invent 'a skirt made in such a manner as to provide the same comfort to the rider as would be experienced if knickerbockers were worn and at the same time secures freedom from accidents which might arise from the skirt catching or being entangled in the pedals or the chain gearing.' The illustration is deceptively simple in comparison to the design. The skirt is made up of single piece of material cut on a bias with a single join and darts for fitting. An Outfitter by trade, John produced and distributed his patented design as *The 'Sandford' Habit Cycling Skirt*. He promoted it himself in advertisements as achieving both 'Elegance and Safety' which made it 'perfect when riding' and 'graceful when walking' and most importantly managed this with 'No Straps, Elastics or Weights'.[13]

A somewhat different take on this theme is Susan Emily Francis's 1897 patent for 'An Improved Cycling Skirt'. Self-identifying as Spinster from the 'Colony of New Zealand', her invention looks like an ordinary A-line skirt from the outside, but differs in that it features a special sewn-in gusset to allow more movement for pedalling legs.[14] She explains: 'It is well known to lady cyclists that the ordinary skirt is uncomfortable when used for riding and by the movement of the legs is unavoidably raised to an undesirable degree.' This garment is more advanced than others within this category, inasmuch as it overlaps with the last theme in this chapter – convertible costumes. This is because it involves switching from one form to another. Emily explains: 'When the skirt is used for walking, the gussets, which would otherwise be unsightly, are hidden by buttoning or otherwise fastening the edges of the gussets together, when the skirt has the appearance of an ordinarily cut skirt.'

Cycling periodicals featured other examples of tailored cycling skirts. *The Lady Cyclist* promoted the 'Thomas', which was a divided skirt designed to conceal the active intentions of the wearer, whether that be cycling or horse riding. It describes how 'at the back the division is hidden in the folds, and in front falls into a slight pleats'.[15] Divided skirts like this were viewed as ideal for 'those wavering between the skirt and the bloomer'.[16] A lady cyclist in *The Cycling World Illustrated* was a convert to this style of cycling skirt: 'For ordinary park riding most ladies wear an ordinary narrow skirt. I have tried various costumes, and generally use the patented design of a West End firm, in which knickerbockers and skirt are combined. It adjusts itself to the machine, and for walking has all the appearance of a plain short dress.'[17]

The prevalence of specifically tailored cycling skirts tells us that they must have appealed to women aware of the dangers of ordinary dress but not interested in looking too much like a cyclist on or off the bicycle. However, this design was not without its critics. A male doctor writing in a column in *The Hub*, 'Should Women Cycle? A Medical View of the Question', was not a supporter of the design. He was reluctant to enter into 'a discussion on the relative merits of the "Rational" or the "Irrational" dress' but nevertheless put a case forward for the knicker-bocker over even a tailored skirt. 'No matter how narrow the skirt may be, there is always the danger of its being caught between the crank and the bracket.' He challenged a 'prominent cycling journal' that had stated that 'such a *contretemps* was impossible' with a story of two women who had 'brand-new, up-to-date cycling skirts' commissioned by quality dressmakers and suffered terrible crashes. 'In one case', he writes, 'the skirt alone suffered; in the other the rider was heavily thrown, and, her head striking the ground, was severely shaken, and was unconscious for some time.' This doctor was in no doubt of the 'risk in wearing a skirt for cycling.'[18]

Built-In Bifurcation

This design solution consisted of a two-in-one design; a skirt combined with a bifurcated garment, such as bloomers or knickerbockers. These inventors had in mind cyclists who were convinced of the practical benefits of bloomers for cycling but not of their aesthetics, or perhaps just simply less keen to place themselves in precarious social situations. This attitude is the kind that fitted with the strategy of concealment discussed in Chapter 3.

For some, wearing this garment might have been an initial step to building up courage to cycle without a skirt. An example of this appears in a short humorous story in *Women and Wheels*, written in 1897, in which the male protagonist seems to have lost his wife to cycling. They had only been married for 18 months when she discovered 'the wheel.' He shares his sorry tale with a friend, and it transpires that it is not only his wife that is missing but also four pairs of trousers. 'They wear them underneath their skirts', he laments, 'but that is only for practice. You mark my words, there will come a day when they will wear them openly. I tell you this thing is interfering with religion.'[19]

For others, the skirt was so central to ideas about femininity that it was not easily budged. In her *Cycling World Illustrated* column 'Cycling

Whispers', Virginia does not celebrate sightings of the 'so-called rational dress' or 'knickerbockered female' in the streets. Although she found the bifurcated garment 'admirable', she was adamant it should be worn 'beneath a skirt'. For Virginia, the skirt was a symbol of woman's 'enoblement' and pleaded with her readers 'for the respect we bear our womankind let the skirt remain'.[20]

Regardless of the motivation, this design solution was a common practice for many and inventors of the time did more than just replace petticoats with bifurcated garments worn under skirts – some sought to combine these garments. An example of a built-in bifurcated design was patented in 1895 by Margaret Albinia Grace Jenkins, a Gentlewoman, of Hyde Park in London.[21] It features an ordinary A-line skirt on a waistband worn over a bifurcated garment with buttoned cuffs. The inventive feature lies in the nature of the combination – the skirt and bifurcated garment are sewn together at the waistband and joined at the sides. Margaret explains that her design 'has for its object to connect the bottom of an ordinary skirt to a pair of knickerbocker breeches worn beneath, as to prevent the skirt rising beyond a certain limit'.

Charles Josiah Ross, an Outfitter in Exeter, patented a similar design in 1897.[22] His invention for 'Improvements in Ladies' Cycling Skirts' similarly incorporates a bifurcated garment inside a long ordinary skirt. However the joining mechanism is different. While Margaret's skirt and knickerbocker shared waist and side seams, Charles' garments are secured at the waist, fork and seat. He explains: 'The wearer will then have on a pair of knickerbockers, the legs of which are secured to and enclosed within a skirt the inside of which is made with a seat for the garment; the parts being so arranged that by the act of sitting down the fullness of the skirt will be drawn forward out of the way of the saddle and the seat of the garment will be in its proper position there being no risk of the skirt hanging on the saddle.'

A slightly different version was made by Samuel Muntus Clapham, a Tailor's Cutter, of Bayswater in London. He patented 'A New or Improved Combined Safety Cycling Skirt and Knickerbockers for Ladies' Wear' in 1896.[23] Here, the knickerbockers are a separate garment and joined to the skirt by a series of hooks on the waistband. The skirt has one side seam sewn shut and the other left open, which can be fastened with

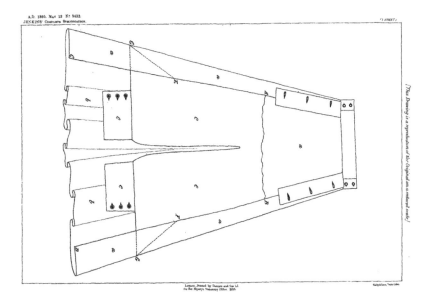

Figure 5.9 Margaret Albinia Grace Jenkin's patent for 'New or Improved Cycling Dress for Ladies'

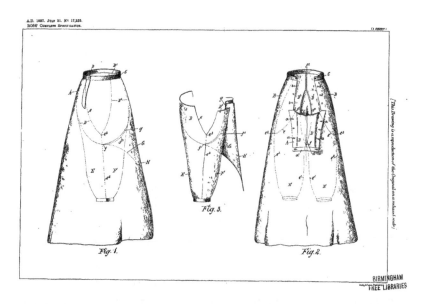

Figure 5.10 Charles Josiah Ross's patent for 'Improvements in Ladies' Cycling Skirts'

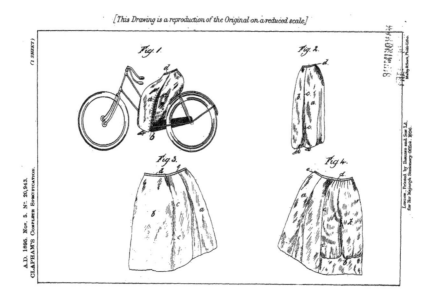

Figure 5.11 Samuel Muntus Clapham's patent for 'A New or Improved Combined Safety Cycling Skirt and Knickerbockers for Ladies' Wear'

hooks so to look like an ordinary garment when walking or away from the bike. It is then unfixed and folded across the front to form a double apron when cycling. This way the excess material is kept away from the wheels while in motion. The convertibility of this patent overlaps with the last category in this chapter.

Bloomers, Breeches and Knickerbockers

Victorian inventors also focused directly on the bifurcated cycling costume. Some of these patented garments were designed to be worn alone and others under skirts and there were many versions; tailored, narrow and very full. This diversity points to the versatility of the garment and also to its unsettled place in late-nineteenth-century Victorian fashion.

Marie Clementine Michelle Baudéan, a Composing Pianist from Paris, focused on a rarely discussed yet critical issue for mobile women who were leaving the comforts of home for longer periods of time – how to conveniently get in and out of garments for the purposes of 'natural needs.' Marie explains how knickerbockers could be complicated to unfasten

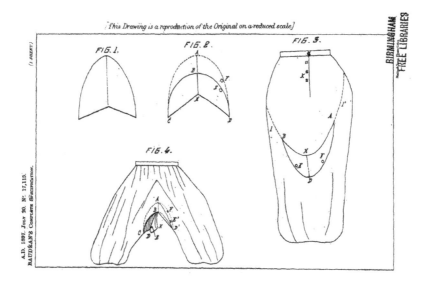

Figure 5.12 Marie Clementine Michelle Baudéan focuses on a removable seat in her patent for 'Improved Knickerbockers Seat with a Movable Side for Female Cyclists, Horsewomen, Huntswomen and the like'

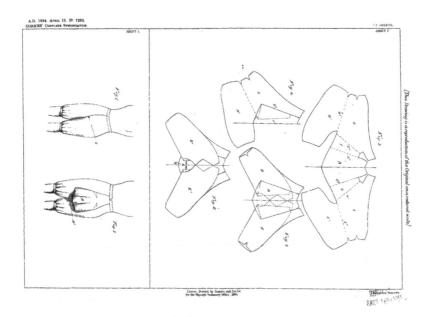

Figure 5.13 James Cornes's patent for 'Improvements in Breeches, Knickerbockers and analogous Garments for Cycling and Riding Purposes'

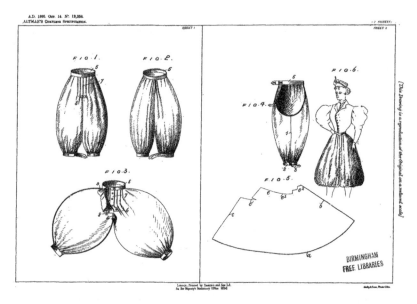

Figure 5.14 Benjamin Altman's patent for 'Improvements in Bloomer Costumes'

under other garments and her design responds with an inventive moveable crotch. 'This invention relates to a new knickerbockers or like seat for female cyclists, horsewomen, huntswomen etc. which will enable the requirements of nature to be satisfied without taking down the knickerbockers or the like or even unbuttoning the waist-strap.'[24]

James Cornes, a Professional Tailor and Cutter of Leicester, addresses a similar concern. He patented a pair of knickerbockers he thought were 'particularly applicable for ladies' use for cycling, riding and similar pursuits'. His patent raises similar issues to Marie in terms of the inconveniences posed by layers of garments 'to comply with certain natural needs'. He responds to this with an invention 'designed to obviate the necessity for so lowering the garment'.[25] Extra material passes between the legs and attaches at the front of the garment, 'which can be opened or turned back to give access to the body for the relief of the person as required'.

Presumably these two inventions were worn under skirts, considering the flexible nature of their seat. Some inventors, however, did away with the skirt entirely. Benjamin Altman, a Merchant, of Madison Avenue

Figure 5.15 Mrs Paul Hardy, *The Lady Cyclist*, 1896

in New York, patented a garment that aimed 'to provide bloomers which, when worn, will closely resemble a skirt.'[26] The sheer fullness of the cut allowed the material to hang in pleats from the waistband as if from a skirt. This type of full bloomer, also known as the Jupe-Culotte, was popular in France 'for the use of those ladies who wish to appear as if they were wearing an ordinary costume.'[27] The voluminous nature of the design made a woman's independent legs difficult to discern off the bike. It also directly addressed problems women had in mounting velocipedes.

Mrs Paul Hardy appears to be wearing these full bloomers in *The Lady Cyclist*'s regular column 'Lady Cyclists at Home' in 1893. She explains in the article how she made this costume herself 'of a light fawn material, made in the French zouave style, finished off with larger pearl buttons, and displaying in the front a soft cream flannel shirt.' The 'zouave style' refers to

The Jupe-culotte.

Figure 5.16 Illustration of a cyclist mounting her velocipede in the jupe-culotte, *The Lady Cyclist*, 1896

the distinctive loose trouser military uniform worn by an elite unit in the French Army in the nineteenth century. Mrs Hardy wore this costume to ride a diamond frame tricycle, which she found 'far safer, swifter and more comfortable than a lady's machine.'[28] *The Rational Dress Gazette* made an array of sewing patterns available for enthusiastic new cyclists. Pattern #1 in April 1899 was for 'Full French Knickerbockers', which it recommended to members because they 'are very full and attract little attention.'[29]

Convertible Cycle Wear

This final design solution, and the focus of the following chapters, presents the most direct and dynamic response to the 'dress problem.' These inventions offered the wearer *convertible* options. They had dual identities; socially acceptable ordinary dress and cycle wear. Wearers were able to switch between the two when needed, depending on the social situation and mobility requirements. These designs gave women choice – to look like a respectable lady, and to change into something more fitting for moving at speed on a velocipede. To do this, inventors built a range of converting systems *into* seams, hems and waistbands of skirts.

Convertible costumes were the most popular form of patented cycle wear during the Victorian bicycle boom. More inventors patented these costumes than any other type between 1890 and 1900, and women designed over half. In fact, convertible cycle wear was the most popular style of patenting by women for women at this time. Inventions in this category are broad ranging. Each features a deliberately concealed system by which the costume changes into a device of mobility – such as weights and pulleys, waxed cords, stitched channels, hooks, loops and buttoning systems. Skirts become capes, bloomers push out of hems and material gathers up out of the way of the wheels. Fundamentally, these inventions did not identify the woman as a cyclist *until* she was ready to cycle, and even then many of them were still not obvious.

Convertible cycle wear differed from rational cycle wear in other ways. Because they were designed to be undetected until the woman approached her bicycle, in outward appearance they were more similar to ordinary women's fashion than rational dress. Rational cycle wear was often made in dark colours, designed to mask dust and mud, and in hardwearing materials like serge and tweed to maximise longevity. Unnecessary frills, ribbons and decoration were limited to avoid things ripping or getting caught in moving parts. This ensured garments were easy to wash and repair when needed, and as a result were often described as 'serviceable' and 'judicious':

> The rational costume in all-wool serge in navy or black comes at 25s., the coat, skirt and knickers (of serge, or black Italian cloth) at 35s. 6d.[30]

> A pretty cycling dress for the country can be made out of brown Holland, either in blouse or coat fashion and with judiciously cut skirt. This sort of thing wears wonderfully well, washes, and does not

show the dust, and will keep down properly if there are little pockets in the skirt hem.[31]

There is no reason why any cyclist should be dowdy. Bright and conspicuous colours should of course be eschewed, but pretty and tasteful costumes can just as well be constructed from dull-toned coloured material. One can wear serviceable clothes without making a dowdy and frumpish appearance ... A badly dressed untidy looking person creates just as much attention, if not more, as a gaudily-attired one.[32]

One of the main aims of convertible cycle wear patents was to avoid looking like a cyclist when away from the bicycle. Not looking out of the 'ordinary' was the central aim. Ordinary clothing at this time for a middle-to upper-class woman meant high quality material, fashionable colours and designs that affirmed her position in society. In this case, standing out in 'bright and conspicuous colours' was fashionable, and paramount to keeping one's cycling intentions hidden. Although not quite as extravagant as high-end visiting dresses, convertible cycle wear would have been closer to fashionable styles than rational dress. Following these rules, convertible inventions deployed fashionability as a tactic of concealment.

Researching, Making and Wearing Convertible Cycle Wear

The rest of the book focuses on and describes in detail five convertible cycle wear patents invented by women for women. The inventors come from Brixton and St. Pancras in London, Maidenhead, Bristol and York, and each patent was registered at the height of the British cycling boom, from 1895 to 1897. Each chapter takes two parts. I start by telling the stories of the women, as much as is possible – where they grew up, family life, education, skills and training – and I look for social, cultural and political influences that gave shape to their ideas. Ultimately, I seek to flesh out their inventions, to bring life to the names written on the patents.

To do this, much like the women who made the costumes, the research weaves together information from an array of sources – from century-old bomb-damaged periodicals to reels of microfiche film. I use data from the Census, Electoral Register, Land Register, Baptism

Records, Marriage, Birth and Death Records, periodicals and newspapers, advertisements and brochures and other commercial materials, personal correspondence, genealogical records and, where possible, contact with extended family members.

Undertaking patent studies is particularly challenging when tracing the history of women inventors. Their names often change through marriage, making them harder to locate at different points. Many conventional research documents of this period are blind to their presence. They were not allowed to vote prior to 1918,[33] so the Electoral Roll is of minimal use in the late nineteenth century for this task. While women have always been listed on the Census, details are sparse. Much of their work for instance goes unrecorded, because it took place in the home and was rarely full-time. Employment such as domestic management and homeworking for family or local businesses likewise is largely unrecognised. Patents similarly tend to omit details of women's vocations. Women may have not considered this an important self-identifier at the time, or data collectors may have been less interested or did not know how to ask for this information. A further common challenge in this field, for both women and men, comes from commercialisation. While this is a marker of success for an inventor, for the researcher it can mean a dead end. Companies often renamed patents, thus making it hard to keep track of the trajectory of specific designs.

As a result, this is not balanced work. As is the nature of research labour, I have discovered much more about some inventors than others. Some women are easier to trace, such as Alice Louisa Bygrave and Frances Henrietta Müller. Others are much more challenging. It turns out there were many Mary Ann Wards in Bristol in the late nineteenth century. Nevertheless, there is more than enough fascinating detail in their patents to explore these inventors and inventions in depth.

The second part of each chapter attends to the making of the inventors' patented garments. Why make them? As we quickly discovered they do not exist. They are not available to look at or touch. We found a plethora of Victorian women's sporting dress in British museums and galleries in the form of yachting, horse riding, gymnastics, swimming and golfing outfits. There were a few late-nineteenth-century cycle outfits but no convertible garments. There are many reasons for this. As any cyclist keenly knows, cycle wear gets worn out. So, these costumes might simply have disappeared through use. Clothing was a precious commodity at the time and it was common practice to reuse material for other purposes. Also, it is important to note that these inventions were deliberately hidden in plain sight. If you did not know what you were looking for, it is likely that the inventiveness of these artefacts would pass unnoticed. Even if examples of these century-old artefacts were available, we (understandably) would not be allowed to try them on, move around or go for a bike ride. So, an obvious reason for making a series of costumes inspired from patents emerges directly from their absence. Another was to think about making and wearing costumes as an inventive method to surface new and different ways of thinking about the past. By making our own versions we had a chance to experience these unique garments in rich, hands-on, bodily practice. We wanted the opportunity to make, touch and wear them, walking and cycling, and generally feel how they both enabled *and* constrained women's freedom of movement.

Although I am adept at sewing, making things on this scale (a total of 27 items) required careful consideration. I am an ethnographer by training, which is where you use your body as a research instrument to gain a deeper understanding of a particular social group or practice. It involves spending time with people, participating, observing and talking, in order to try to see the world through their eyes.[34] Doing ethnography is, of course, not possible when your research subjects lived over a hundred years ago. So, we chose to spend time with the traces of their lives. We observed, participated and examined the women's lives through their patents and the process of making the garments.

Undertaking this kind of practice research archival work involves creativity, imagination and energy. It goes beyond a distinction of research and bodies, them and us, and then and now. One of the aims of the project was to explore what might emerge from making that differed from just reading and textually analysing the patents. In many cases, what we felt was a deep appreciation of the complexity of these seemingly simple garments. They looked ordinary from the outside, which was their intention. But in many cases the complexity of the design lay inside the garments, which only really emerged when we started to make them. Each had to be made up to four times; small-scale, toile (full-scale version in a light material), mock-up (full-scale version in a similar weighted fabric to the final) and final garment. In (re)making these pieces, the intricacies in even the simplest details revealed themselves.[35] Buttonholes, for example, were not always enrolled for the purposes of buttoning. They could be, in the case of Alice Bygrave's patent in the next chapter, a critical component in a pulley system concealed in the seams of a skirt. While the making section forms only a small part in each chapter, it was formative to the articulation of ideas and the piecing together of both the garments and the stories.

Overall, the following biographical chapters are underpinned by the desire to remember and reclaim an important part of British women's cycling history. The reality of writing from multi-dimensional artefacts – patents, costumes, archived periodicals, genealogical records, personal correspondence – is to recognise multiple and partial accounts, to acknowledge gaps and overlaps, and resist the desire to tidy up and erase ambiguity. My navigation through these many materials involves stitching some parts together while unpicking others. It is a deliberately messy intervention in the smooth narratives of technological history that has concealed them until now. While this method of sense-making was amazingly generative, it also proved difficult to know when to stop. Much like a treasure hunt, shiny things kept appearing, layering and building thicker and richer stories, even while I was writing them, that gave me pause to rework and re-configure the narratives. Critically, what the nature of these dynamic socio-historical materials remind us is that we can never capture or fully know the entire story and that there were many stories yet to tell. This is just the start.

Part II

6

Patent No. 17,145: Alice Bygrave and Her 'Bygrave Convertible Skirt'

Alice Louisa Bygrave submitted a complete specification for the British patent 'Improvements in Ladies' Cycling Skirts' on 1 November 1895. It was accepted that same year, on 6 December.[1] The patent tells us that she was a 'Dressmaker' and was living at the time at No. 13 Canterbury Road, Brixton, in the county of Surrey, South London.

Like many cycle wear inventors of this period, Alice set out to find a way to cycle safely and comfortably, yet also appear less like a cyclist when away from the bicycle. 'My invention', she declared, 'relates to improvements in ladies' cycling skirts and the object is to provide a skirt as proper for wear when the wearer is on her cycle as when she has dismounted.' The use of the word 'proper' is revealing of the pressure placed on women and their clothing to be practical *and* feminine, even when these demands seemed at odds in differently socio-mobile and technological contexts. How does she do this? Alice *improves* an 'ordinary skirt' with a series of hidden devices sewn into its infrastructure. Her design is the most mechanically oriented of the collection. It features a carefully concealed system of buttonholes, weighted hems, stitched channels and waxed cords that together comprise two pulley systems, sewn into the front and rear central seams of the skirt. Working with materials at hand (skirt, cords, buttonholes) she sets out to expand and re-configure what they do – this is a classic hacking approach. It produces a unique system, which enables the wearer to adapt their costume when needed. She explains: 'As the wearer prepares to mount her machine, she pulls both cords in from the top, thereby raising the skirt before and behind to a sufficient length.' The pulleys gather the material, much like a curtain, and when the cords are affixed at the waist, hold it draped over the hips in a fashionable ruche style of the time. This efficiently lifts the skirt up and out of the way of the wheels, pedals and moving legs. The weighted hem then permits a quick reverse action. When the cords at the waist are released, the gathered skirt material drops and the wearer quickly conceals both her lower limbs and her cycling intentions.

A rich theme throughout Alice's story concerns the collaborative nature of her design and ideas. Although she is listed as the sole named patentee, which remains the usual convention, there is ample evidence of a plethora of influences. Her inventive response comes from her dressmaking skills, cycling interests and exposure to her family's watch- and clock-making business as well as the diverse range of crafts people who moved through her life. Before looking closely at the skirt it is useful to understand more about the life Alice and her family led and also trace the journey of the patent. It turns out that her convertible cycle wear costume not only helped women move in new ways but it also ended up travelling great distances.

The Inventor and Her Life

Alice must have known that she was onto something exciting and potentially profitable with her improved cycling skirt because this British patent was just the first of several that she registered. Ambitiously, she applied for three more international patents for the same design, two within a few days of each other and another a month later. On 13 December 1895 she claimed her design with the Swiss Federal Institute of Intellectual Property and on 18 December she had another to her name at the Canadian Intellectual Property Office. Although formatted differently, they outline the same invention. The Canadian document is particularly interesting as it features hand-written corrections to the typed description and bears Alice's signature. Alice lodged a further version of her design with the United

N° 17,145   A.D. 1895

Date of Application; 13th Sept., 1895
Complete Specification Left, 1st Nov., 1895—Accepted, 6th Dec., 1895

PROVISIONAL SPECIFICATION.

Improvements in Ladies' Cycling Skirts.

I, ALICE LOUISA BYGRAVE of No. 13 Canterbury Road, Brixton in the County of Surrey, Dressmaker; do hereby declare the nature of this invention to be as follows:—

My invention relates to improvements in ladies' cycling skirts and the object
5 of it is to provide a skirt proper for wear when either on or off the machine.

In carrying the invention into effect, I make use of an ordinary skirt and the ordinary knickerbockers, the former being worn over the latter. I fasten a tape or cord to the bottom edge of the skirt in front and carry this cord up through suitable guides to the top of the skirt where it is made fast in any convenient
10 way. The back of the skirt is fitted with a cord in the same way as the front.

As the wearer prepares to mount her machine, she pulls both cords in from the top, thereby raising the skirt before and behind to a sufficient heighth, after which she makes the cords fast in any convenient way, as for instance to the same fastening to which the top ends of them are held. This raising of the skirt
15 before and behind leaves the sides of it festooned, as it were, over the knickers. Before dismounting, the above mentioned cords are loosened and the skirt pulled down, back and front, by suitable weights made fast to the bottom edge of it.

Dated this 13th day of September 1895.

PHILLIPS & LEIGH,
20   Agents for the Applicant.

Figure 6.1 Excerpt from Alice's 1895 cycling skirt patent

Figure 6.2 Illustration from Alice's 1895 cycling skirt patent

States Patent Office two months later, in February 1896 (and there is evidence that the costume also travelled to Australia). With these four patents she effectively laid claim to her invention across several major continents.

Why did she do this? As discussed in Chapter 4, the inventing fervor of the late nineteenth century must have been infectious for many parts of society. A broader range of people were being encouraged to patent their ideas, which made a refreshing change given barriers to entry into this world had long been high. There were many possible reasons why Alice sought to patent her design so widely. She could have been reading publications like *The Queen* that discussed the challenges and opportunities of 'Women as Patentees'.[2] Similarly, publications like *The Dawn* and *The Women's Penny Paper* presented different ways women could develop their skills and get involved in business. She could also have been wary of having her ideas copied and stolen, as stories of lost wealth through theft and corruption were just as rife in the newspapers as those of commercial triumph.[3] More catalysts undoubtedly came from her home life.

Alice identifies herself in the patent as a dressmaker, probably having learned from her mother. Emma (née Reed) was born 1834 in Covent Garden in London and worked as a dressmaker. She married Alice's father, Charles William Rudolph Duerre, in 1854 at The Strand. Unlike Emma, Charles was not a Londoner. He was born in 1830 in Memel Prussia and worked as a watch- and clock-maker. Alice was born on 28 July 1860, and she had two older siblings (Emma and Rudolph) and five younger ones (Arthur, Ada, Earnest, Edward and Bertha). Living in a family of this size would have made for a hive of activity, but there was even more bustle in the Duerre household. According to the 1871 Census, when Alice was 11 years old, the Duerre's were living on Kings Road in Chelsea in West London. The faded hand-written document tells us that on the night the record was taken, 2 April, the house was inhabited not only by the immediate family, but a further seven people. Either the Duerre residence was a boarding house or the family resided in a shared house with fellow craftspeople. The former seems more likely as the records list these people as 'lodgers'. The age and occupations of the lodgers add further texture to Alice's story. They were Henry Sayers, 28 (Artist Brushmaker) and his wife Ameila Sayers, 27 (Artist Brushmaker's wife) from Marylebone and the Cooper family – Thomas E. Cooper, 22 (Dyer and Cleaner), Louisa C. Cooper, 26 (Dyers Shopkeeper), Emily C. Cooper, 25 (Dress Maker), Ebenezer, S. Cooper, 21 (Watchmaker) and Henry C. Cooper, 29 (Dyers Finisher), all of whom came from Islington.[4] Alice would have grown up surrounded by a group of young people engaged in creative practical occupations.

The larger context of Alice's early life was no less animated. Her parents owned and ran a watch- and clock-making shop, also on Kings Road in West London. A photograph of it reveals a window teeming with meticulously arranged goods. Rows of tiny, round clock faces off-set grand carved wooden timepieces. Tiered display cases hold pocket watches and fobs of all kinds and pendants swing on long chains. The window is framed with ornate stained glass, which would have further enticed passers-by to pause. Time played a central role in Victorian culture. London's Big Ben was completed in 1859 and the increased availability of personal watches and clocks symbolised a commitment to industrialised working practices and new mobility technologies, such as the railway. In fact, the combination of invention and time (along with cycling) inspired many great works of fiction, such as H. G. Wells' *The Time-Machine*, written in 1895.[5]

Given it was a watch- and clock-*making* shop we can imagine that inside was partly given over to a workshop for restoration and repair. Walls and benches were probably similarly clad in goods for sale, along with tools of every description – springs and cogs, pin levers, ornate winding keys, carved hands and painted dials, wheels and gear systems of all scales and metals. Here, Charles's and Emma's expertise and knowledge would have been materially displayed. The crammed window display suggests little natural light would have filtered into the shop. With electric lighting still in its infancy at the turn of the century, it is likely gas lamps would have punctuated the dim interior with glimmering light. Entering the shop might have felt like stepping *into* the intricate anatomy of a timepiece, into the very mechanics of its secret innards. The sounds too would have been very different inside from those of the

Figure 6.3 Our version of Alice's patented convertible skirt in action

streetscape; at the threshold, the noise of the street would have given way to a cacophony of clicking ticking, chimes and bells.

Outside, Chelsea was a hub of industry, buzzing with goods and people. Kings Road was so named because it was King Charles II's private road to Kew until 1830. The nineteenth century saw it expand to become synonymous with art, textiles and design and was home to, amongst others, J. M. W. Turner, Virgina Woolf, Oscar Wilde and the Chelsea School of Art. The *1895 Post Office London Directory* provides us with a sense of the surrounding shops and cafes at the time Alice's family lived and worked in the area. There were linendrapers, bootmakers, hosiers, dressmakers, upholsterers, tailors, hatters, hairdressers, butchers, greengrocers and fruiterers, wine merchants, confectioners, fishmongers, cheesemongers, leather sellers, chemists, surgeons, stationers and even a cyclemaker.[6]

In 1881, at the age of 21, Alice married Charles Edward Bygrave, a commercial clerk for an International Sleeping Car Company (passenger trains with sleeping berths) and together they moved to No. 13 (also referenced as 13–15) Canterbury Road, Brixton. They had two children, Lena Alice in 1883 and Herbert Charles three years later.[7] The ambiguity of the house number is relevant as the 1891 Census reveals that in addition to the family, on 5 April, there were five other adults in residence.[8] Much like how she was raised, Alice continued to live in a boarding house.[9] However, interestingly, things were changing from her parents' generation. Not everyone in the house was local. The Census accounts for a 'visitor', an Alice Maud Lucy Foskett, 22, from Kentish Town and a 'General servant', Florence Elizabeth Chandler, 18, originally from Stevenage. Three 'boarders' were not English, indicating an increase in workers from abroad at the time. Edmond Herbert Hippolyte Fletcher, 20, Alexandre Debisschop, 20, and Fernand Pequewand, 22, were all listed as wool buyers from different parts of France. Alice's patented design might in part have been inspired by her exposure to people from different parts of the world directly engaged in hands-on craft, engineering and textile vocations.

Bicycle culture and the location of Alice's childhood are also central to the telling of her story. Kings Road in Chelsea would have been a thoroughfare of cycling activity on a daily basis and especially so when it was the site of cycling parades. Tessie Reynolds's record for cycling from Brighton to London and back, for instance, was the catalyst for one such event in October 1893. The streets were bustling with excitement about her endurance feat and what she had chosen to wear to undertake the ride. Both generated considerable interest as per this writer's account in *Bicycling News*:

> By her ride, and the extraordinary correspondence it has provoked, together with the reproduction of her photograph throughout the country, Miss Reynolds has accomplished more in three weeks in stirring up opinion about ladies' rational dress than could otherwise have been achieved in as many years. In Brighton drawing rooms and at dinner tables allusions to the subject are frequently heard. Last Sunday, accompanied by an escort, Miss Reynolds rode along the King's Road, when the sidewalks were crowded with promenaders, including our informant, who states that every comment he overheard was favourable and friendly, both as regards the rider and her dress.[10]

Given the volume of attention, it is unlikely that Alice could have missed the spectacle passing along in front of her parents' shop and may well have been on the street, in amongst the crowd. Even if she had not been there, she might have heard from family members or read about it in extensive media coverage and may have taken inspiration from Tessie's costume. There was of course a flip side to the mass attention. Although generally positive, this writer in 'The Ladies' Page' of *Bicycling News* explains her misgivings: 'I hear that Miss Reynolds, who wears a French costume at Brighton, attracts an immensity of attention, and that her photographs, which are for sale in the shop windows at Brighton, are in great demand. She is very young – only about fifteen – and probably does not mind the attention she attracts as much as she would do if she were older.'[11] At 33, and comparably an older woman at the time, Alice might have also found the attention surrounding Tessie's costume overwhelming and invasive. It is possible that both feelings, excitement and wariness, were formative in shaping her material response to the 'dress problem.'

Further inspiration for Alice's design may have come from her parents. Her father was an everyday cyclist. He used his bicycle to collect and deliver watch- and clock-making goods around London. Apart from practical use he indulged a keen interest in bicycle design, particularly in saddle construction. Like his daughter, Charles Duerre was also a patent holder. In total, he laid claim to five inventions relating to the bicycle and the timing of each is interesting. Charles submitted his first British

Figure 6.4 The Duerre family's watch- and clock- making shop on Kings Road, Chelsea, in West London

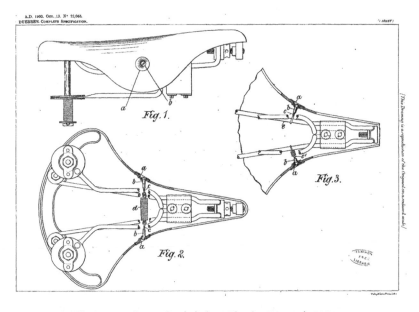

Figure 6.5 Illustration from Alice's father, Charles Duerre's 1904 patent

patent in 1894 for 'Improvements in Cycle Saddle Springs', a year *before* Alice's British patent.[12] Alice may have become familiar with the patenting process through his experience. Perhaps she spent time with him in the shop talking about ideas and making models. Charles continued to develop his ideas over the next decade. He submitted an improved version of the invention in 1899 and shortly afterwards followed his daughter in lodging the design with the Austrian Patent Office in 1900.[13] Maybe here, the tables were turned and father learned from daughter. Charles continued to develop his saddle designs with further UK patents in 1901 and again in 1904.[14]

Looking closely at Charles' inventions provides insights into Alice's approach to cycle wear. Knowing that he was a trained watch- and clockmaker goes some way to explain his interests in getting *inside* the saddle, under the leather surface and into the structural framework of springs, coils and rivets, to set about reworking the principles of suspension and reconfiguring the tensile forces at play. It is appropriate that Charles approached the saddle much like he would a broken watch or clock: opening up the device; dismantling many different pieces that work in complex interconnectivity under the surface; seeing it as a sum of parts rather than a seamless coherent whole. Combined with skills learnt from her dressmaking mother, this goes some way to explain why Alice came to engineer complex pulleys, technical systems and devices *into* the infrastructure of her clothing and how her invention was purposely hidden in plain sight, under the surface – much like the cogs in a watch.

It also makes sense that both Alice and her father used the same patent agents to submit their completed specifications to the Patent Office. Phillips & Leigh Patent Office was officially founded in 1882, but has records dating back to 1876. It was clearly supportive of not only small independent traders, but also of an even more marginalised subset, women inventors. It continues to operate today, over 130 years later, out of a London office near Chancery Lane and the Inns of Court (a short distance from its original site at No. 22 Southampton Buildings). The Patent Office Library was also located in Southampton Buildings and was apparently well used by hopeful inventors with the highest number of readers recorded in 1897, during the height of the cycling craze.[15] Much like technology hubs today, in the 1890s this part of London was home to a dynamic cluster of patent agents, information centres and supporting trades.

Another significant influence in Alice's life came from her younger brother, Arthur, and his wife, Amelia Rosina, who were both professional racing cyclists.[16] Although her married name was Duerre, Amelia used her maiden name to race – Rosina Lane. As discussed earlier, cycle racing was immensely popular in the 1890s, predominantly for men. The thrill and excitement of this new sport produced and reinforced its masculine appeal and the cycle industry greatly benefited from this association (as did everyday male cyclists who could purchase the newest cutting-edge technologies). The culture of racing trickled down into urban cycling and reinforced cycling's 'fit' with many masculine ways of being in public space, including producing in some instances anti-social behaviour in the form of 'Yelling Yahoos', 'Cads on Castors' and 'Scorchers'.[17] In contrast, women were expected to cycle with 'grace' and 'modesty' – a cultural discourse that did not translate well into women's racing.

Yet, race women did. And it was an immensely controversial and popular subject of debate and discussion at all levels of society. Their presence at big racing events was largely constructed (and controlled) as 'novelty' athletes and routinely overlooked in contrast to 'proper' (male) athletes and record-breakers. It is only recently that scholars have been (re)writing active women (back) into cycling's history. Fiona Kinsey, for instance, set out to 're-insert female Australian competitive cyclists into the historical record' in her research into Australian cycle-racing women in the 1890s.[18] Similarly, the history of British women's racing is the focus of Clare Simpson's work. She writes about how 'on the public streets, women were strongly sanctioned from "fast and furious riding", and so it was a rare thing to witness such a phenomenon and for any duration'.[19] Part of the attraction of participating in and watching women's racing was the excitement of actually seeing women doing fast, daring cycling. However even at dedicated racing sites such as an indoor velodromes like the Royal Aquarium in West London where Rosina regularly raced, the media begrudgingly covered women's events. 'To a man of old-fashioned views it is rather a *bouleversement* [upset] of ideas to watch perspiring females cycling against time for all the world to see.'[20] Another wrote: 'It must be conceded that woman – lovely or otherwise does not appear to great advantage in a bicycle race.'[21] A columnist in *The Queen* expressed amusement at this attitude: 'The cycling press has almost unanimously expressed strong disapproval, and in one or two cases the writers have been almost hysterical in their expressions of disgust.' In contrast, this writer saw little wrong with it and even acknowledged 'that some women will race as a means of livelihood.'[22]

Rosina Lane did exactly that. She was able to race enough to earn £100 per annum for five years as a professional cyclist at Olympia and the Royal Aquarium. This was a significant sum, around £12,000 in today's money. Because she regularly placed well in events, her name and image often appeared in newspapers. What is all the more remarkable about Rosina is that she had four children, all daughters, before her racing career really got started.[23] *The Sketch* called her 'a most improving cyclist' in 1896, in a report about how she had cycled from London to Brighton and back 'in fairly good time' and how her 'chief successes were in handicaps at the Aquarium, where she won twenty and two twelve mile races.'[24] She was apparently also good at endurance racing. On Sunday 6 December 1896, a London newspaper reported on the 'Finish of Novel Ladies' Cycle Race' at the Royal Aquarium.[25] This six-day event involved 'four teams of three riders each wheeling eight hours per day'. The competition was made extra challenging for the captains, as each team was 'handicapped' with a slow rider. It was a close race. Rosina led the Yellow Team, which may have scored the lowest overall tally, but her personal distance was an epic 365 miles and 5 laps, and the fourth highest out of the 12 riders.[26]

Being related to a successful female racing cyclist must have generated lots of discussion at family get-togethers and potentially inspired Alice to design her cycling skirt. Another reason that Rosina Lane's racing life is relevant to the story is because she was photographed wearing her sister-in-law's costume. In fact, she is pictured wearing the costume *before* it was patented. On 27 November 1895, she appeared in *The Sketch* standing next to a diamond frame bicycle, which was most likely her personal racing machine. She looks comfortable and confident and wears the garment so well it looks tailor-made for her – which it probably was. In other portraits she is positioned in the centre of a group of racing cyclists. She photographs well. Her costume stands out against all the others because she is the tallest cyclist and centrally positioned in the group. It is also because of the bright, light-coloured fabric. It's not implausible that this was planned. These articles do not mention the costume but it is vastly different to all the other racers' outfits. It almost certainly would have drawn comment. As we discover later, Alice was not only an inventor but also an astute businesswoman. Rosina's physical presence, cycling prowess and media appeal would have made her a perfect role model to promote Alice's invention.[27]

Figure 6.6 Miss Rosina Lane, dressed in Alice's patented cycling skirt, is one of 'The Lady Cyclists at the Aquarium', *The Sketch*, 1895

Figure 6.7 Rosina Lane dressed in Alice's patented cycling skirt (centre),
The Sketch, 1895

Sadly, Rosina's story does not end well. She was involved in a terrible crash that put a devastating end to her cycle racing career. On 21
August 1897 she was knocked from her bike in Richmond Park, south-
west London, by a horse and carriage. Her body and bike were pinned
between a bridge wall and the vehicle. The bridge gave way and all
tumbled into the dip below. It was a serious incident with tragic consequences. A court report, two years later, provides more details of the
terrible event: 'The bicycle was smashed and the "rational costume" in
which she was riding was torn to pieces.' The driver and carriage wit-
nesses claimed that she hit the horse as 'her steering was somewhat
erratic' and she was 'riding fast and bending right over her handle-bars.'
The opposing lawyer brought up the fact that she was a well-known racer
and had once been fined 10s for riding fast in Kingston. He was clearly
trying to attribute some of the blame to the victim, by suggesting her vocation was partly responsible. The judge disagreed, noting evidence that
showed the horse and carriage were on the wrong side of the road when
the crash happened. Rosina won the case but was awarded a paltry £55
in damages for her suffering.[28] She survived it but remained in pain, no

longer able to race, or even to cycle. She was not even able to manage the family's Kings Road shop with her husband.[29] Rosina's cycling career was bright but sadly all too brief.

Commercialisation and Distribution

Patents are important repositories of information regardless of whether they reach the heights of commercial success or remain in the archive. As discussed in previous chapters, these records provide valuable glimpses of life in previous times. We gain insights into how inventors were imagining intersections of bodies, social norms and technological advances. To trace ideas from descriptions and drawings through the process of production and finished material objects, to how they are promoted in media across different countries and end up in the hands (and on the bodies) of eager consumers, is the researcher's dream. For

Figure 6.8 Rosina Lane (left) and fellow racers Miss Murray and Miss Blackburn on a triplet, *The Sketch*, 1895

Figure 6.9 Jaeger advert for the Bygrave 'Convertible' Patented Cycle Skirt, *The Lady Cyclist*, 1896

many, however, it is all too rare. The reality is more frequently a dead end. Often this happens when a commercial producer renames a patent. The process can be abruptly cut short in even more ways for women, whose contributions have conventionally garnered less public awareness, celebration or even mention in conventional records.

This is not the case with Alice's patent. In fact, it is the opposite. 1896 proved to be a very big year for this new inventor, then 35 years old and mother of two small children, as her patenting activities took off. Incredibly, her invention was produced, commercialised and distributed by Dr Jaeger's Sanitary Woollen System Company Ltd. The Jaegar Company was a substantial backer for a new patentee. Established in Britain in the early 1880s by Lewis Tomalin, and inspired by the charismatic German scientist Dr Gustav Jaegar, the company promoted the sanitary benefits of high-quality natural fibres worn close to the skin. It was deemed particularly suitable for active wear and highly regarded by many, including The Rational Dress Society, which wrote positively of the company's goods in the first edition of *The Rational Dress Gazette* in 1899: 'The Jaegar Company, Regent St., are now making all wool knickerbockers, suitable for outdoor wear, at a moderate price. They are a very good pattern.'[30]

Jaeger held the sole British licence of what they called the 'Bygrave "Convertible" Skirt'. They produced the skirt in a range of Jaeger fabrics and sold variations at different prices. They also promoted the new design in popular women's press and cycling periodicals. A full-page advertisement for the 'Bygrave "Convertible" Skirt' appeared in *The Lady Cyclist* in March 1896. It promoted the skirt as 'patented', which was a highly valuable selling point, and specifically designed for 'Cycling and Walking'. The speed of the convertibility is stressed with the claim: 'Instantaneously raised or dropped'. The copy is accompanied by two illustrations of these benefits in action. One depicts a woman on a bicycle pedalling with the skirt raised out of the wheels. The other shows her walking next to her bike with the skirt completely down and giving no indication that it is anything other than an ordinary skirt.

Many journalists recognised how Alice's skirt operated as a dynamic medium *between* fashionable and rational clothing and discussed at length how it occupied this liminal space. *The Queen* reviewed the 'Bygrave "Quick Change" Cycling Skirt' and heralded it as a 'happy solution to the vexed question of ladies' wheeling dress', because it 'hits the golden mean between the ordinary and the rational, giving to the rider all the comfort of the latter, and the additional ease of knowing that in a moment it can be resolved into a perfectly ordinary skirt'.[31] The writer made further positive comment about the reduced volume (only

two-and-a-half yards) of material, which meant 'there is no unnecessary fullness over the hips'. Buyers could choose from a range of 'durable,
unweighty and pliant' materials from the Jaeger natural fibre collection.
Skirts could also be made with and without lining. These choices meant
the tailored garment could be customised to fit taste and budget.

Alice's invention was promoted at popular events such as the Stanley
Cycle Show at the Agricultural Hall in Islington. This was a much anticipated and reviewed annual exhibition of everything new in the world
of bicycles, accessories, machinery and horseless carriages (new motor
cars). The costume was showcased at least twice, in 1895 and again in
1896, in a promotional stall run by Jaeger. The convention for selling
bicycles at the time involved inviting people to ride machines prior to
purchase. Jaeger did similar with cycle wear. It encouraged people to
see Alice's costume in action. Adverts attracted passers-by to the stand
where staff were 'demonstrating the very obvious advantages of their
new cycling skirt'. While the company was promoting all kinds of active
wear, from shoes and stockings to suits and waistcoats, the only specifically named garment was the 'Bygrave' convertible skirt.[32]

The Stanley Cycle Shows ran from 1878 to 1911, but the 1896 show
was considered to be at the height of cycling's popularity. *The Lady
Cyclist*, in March 1896, makes favourable comments about Jaeger's overall contributions to women's cycling and draws particular attention to
the convertible skirt. 'The "Bygrave" costume which was exhibited at
the Stanley Show, is designed for either cycling or mountaineering.'[33]
Perhaps the writer was struck by the possibility of this unusual combination of activities after seeing a demonstration of the costume. They were
not alone. A writer from the professional tailoring periodical, *The Tailor
and Cutter*, as well as several onlookers, was also clearly taken by it:

> The 'Bygrave' skirt is shown at another stand, and is the invention of
> a Brixton lady ... The stand was the centre of an admiring group of
> gentlemen, who watched with interest the two lady models convert-
> ing the skirts they were wearing into the bloomer-like knickers, by the
> simple operation of pulling the two cords in the front of the waist.[34]

The Tailor and Cutter writer reviews the rest of the large Jaegar
exhibit featuring 'all kinds of garments which may be used in cycling'.

What is particularly interesting is how they make a distinction between Alice's skirt and examples of rational dress on the stand. While Alice's design was clearly a more *rational* choice for cycling than ordinary skirts, it was not labelled as rational dress. It was called the 'Bygrave', which firmly associated it with a female inventor and live demonstrations by 'lady models' clearly did not do it any harm. This was an astute business move by Jaeger, given the socially polarising response to rational dress. Carving out a new space between divisive categories would have attracted a broader audience and market.

What clearly piqued people's interest was the flexibility Alice's costume offered women cyclists to imagine new ways of being in and moving, without abuse or harassment, through public space. A critical advantage for a writer in *The Wheel Woman and Society Cycling News* was how the garment transformed, meaning the wearer was not fixed to one identity, but rather was free to choose between walking and cycling.

> At Dr Jaeger's Stall I was much taken with the 'Bygrave' safety cycling skirt, which I was informed '*Does not blow out, catch in pedals, or ride up, and can be instantly converted into "Rational Costume"*'. All of this I can endorse, having seen it. When down for walking, it has the appearance of a neat and nicely-cut skirt, and is raised for riding by simply drawing two cords; on releasing the cords the skirt reverts to the usual shape. The skirt, complete, can be purchased from one guinea, which brings it within the reach of all; knickerbockers to match from 7s. 6ds.[35]

The combination of inventiveness, flexibility and quality appealed to people. Cost was also a considerable selling point. At this time, a pound was made up of 20 shillings or 240 pennies. So, the price of one guinea meant the skirt was sold for one pound and one shilling (or 21 shillings). According to this writer, it was considered 'in reach of all'. Was it? To provide context we can examine average Victorian wages. Women earned less than men, but a head nurse might take 25 pounds home every year, a lady's maid's wage might be up to 30 pounds and a housekeeper could earn around 50 pounds.[36] (It was also much cheaper than a custom made cycling suit, such as those promoted in the *Rational Dress Gazette*.) Clearly

SLIGHTLY RAISED. DOWN FOR WALKING

Figure 6.10 Illustrations of the Bygrave Skirt in action in *The Westminster Budget*, 1896

higher-class women had more disposable income, so it seems likely that the garment was indeed priced for a broad consumer market.

Alice's costume seemed to address an identified gap in the market. There was a desire for good quality, affordable and fashionable cycle wear for women. Patented goods also held an attraction for the general public. The skirt certainly found favour with the *The Westminster Budget*, which spent column inches lamenting the lack of appropriate garments until the discovery of the Bygrave skirt:

> But at last it seems that a skirt has been brought out which is, in every way, the right thing to wear at the wheel; and I am glad to see that the Jaeger Company are bringing out this skirt, for we have all learnt, in the course of a good many years, to put implicit faith, if not in *all* the theories of Dr. Jaeger's Sanitary Woollen System Company,

MRS. BYGRAVE'S BICYCLE SKIRT.

Figure 6.11 'Mrs. Bygrave's Bicycle Skirt' is illustrated in the *St Louis Post Dispatch*, 1896

at all events in every article of clothing they supply. The Jaeger materials for daintiness, simple elegance, and durability are second to no other stuffs, and they are sanitary besides ... The skirt in question is pure wool, as all cycling skirts ought to be; it looks most fashionable and neat when its wearer is off the wheel. Then, by drawing two magic and invisible cords, Mrs Bygrave, the inventor, causes the skirt to be raised to any height required while riding, and you ride as comfortably and look as nice about your skirts as the most fastidious could desire. And the Bygrave skirt has wisely been put into the market at the reasonable price of a guinea.[37]

In addition to selling from its own flagship depots in Central London and attending popular cycling events, the Jaeger Company plied the

Bygrave skirt to its many agents and distributors. These included local companies, such as silk merchants and general drapers like E. Broad & Son, in Aylesbury and Cooper Hunter & Rodger, in Glasgow. Like many smaller companies around the country, they were agents for Dr Jaegar goods and exploited the growing popularity of ' "The New" Bygrave Convertible Skirt' to draw customers into their stores.

The invention travelled even further (perhaps Alice's husband, Charles, in his job as a commercial clerk in an international travel company, was involved in some way). Unlike the Swiss and Canadian patents, we know much more about the drama surrounding Alice's patent in the United States Patent Office on 25 February 1896. An American Sunday morning broadsheet, the *St Louis Post Dispatched*, thrilled its readers on 8 March with a story about a new 'rapid and exciting business transaction' by a 'young and pretty British Matron' that 'made $5000 quickly'! Alice had apparently travelled from Britain to New York at the end of January to promote 'a bicycle skirt of her own invention'. The journalist reports that within two hours of arriving on American soil, 'while still nervous and dizzy' from the journey, she 'not only donned her new original skirt, but was displaying its various merits to the buyer of one of the largest sporting goods establishments in the country'. However, Alice did not accept this first offer and instead went to a competitor where she did an even better deal. Alice's business acumen clearly impressed witnesses:

> Not satisfied with the offer he made her, this self-confident, energetic young woman went forth in search of greater financial inducements to part with her cycling skirt. The first New York establishment she visited offered her a royalty on all sales made. But Mrs. Bygrave had other ambitions and walked away in her patent skirt to a well known firm on twenty-third street. The wisdom of her choice was shown when she promptly received an offer of $5,000 from the Twenty-third street dealers.

This was a vast sum of money. In today's terms it translates to nearly US$130,000. There was even more to come. The journalist continues to build the drama. Apparently the deal was negotiated on the basis that Alice could get a US patent, which at that stage she did not have. So, she immediately set off to Washington and spent two hours in the Patent Office to get this in motion. She then returned to New York to confirm that this was possible and on the Wednesday, four days after she had arrived, she sailed home. The paper was clearly an enthusiastic supporter of her ambition and design. In fact, it describes the invention in more detail than any others until now. Perhaps Alice had been developing her sales pitch and demonstration, as the skirt seemed to have even more convertible options.

> Her cycling skirt is the most novel invention in the matter of wheeling skirts that has yet come before the wheeling public. By a system of cords worked through openings near the waistline it can be made to fill three different varieties of long-felt wants of the bicycle woman.

The first of these three variations was the straight walking skirt. The second involved the gather or 'shirring' of the front pulley system to just above the knees. This, according to the journalist, adapted the skirt for drop-frame cycling. The third version involved the use of both sets of cords at the front and back of the skirt for use on a diamond frame, which interestingly is how Rosina Lane is shown wearing the costume.

> The third possibility of this unique garment is developed by working the cords that run up the back seam and find an outlet under a tailor-made flap just over the hip. The pulling of these two side cords converts the skirt into a pair of neat and graceful bloomers that will permit the fair rider, in case of emergency, to mount a diamond frame bicycle with all the grace and utility of a masculine wheelman.

The costume continued to generate enthusiastic coverage in the US media. It was covered again in in the Sunday edition of *The Saint Paul Daily Globe* in March with the headline '$5000 in Four Days: Mrs Bygrave received it last week for a bicycle skirt.'[38] The next month *The San Francisco Chronicle* featured it in a section called 'For The Girl Who Would A-Wheeling Go', which covered current fashions and materials for keen cyclists. Next to an illustration of the converted Bygrave costume it discussed the issue of how many women cyclists would often cycle in bloomers but usually with skirts and a petticoat over the top. This was cumbersome, so some removed these outer layers, wrapped them up and attached them to handlebars when no one was looking. (This is the premise for the patent in Chapter 9.) The writer saw the Bygrave skirt as a more advanced convertible option:

> A more convenient arrangement than this is the new Bygrave skirt, named after its inventor, who is an English woman. The skirt is the product of her own experience and is very simple, yet very effective. The idea was to arrange the skirt in such as manner as not to interfere with the free management of the pedals, and to prevent its catching on the wheels. The skirt is practically converted into a pair of bloomers by drawstrings running up and down the middle of the front and back of the skirt. These strings many be pulled as tight as is desired, raising or lowering the skirt at will, and they are provided with catches to hold them in place. The skirt may thus become a pair of knee bloomers, or be allowed to hang loose like a divided skirt: and when worn amid 'the busy haunts of men' it appears as a plain, ordinary skirt, with never a suspicion of masculinity about it.[39]

This might be a case of journalists waxing lyrical, but there are clues in these articles that help to explain why Alice's invention remained in her name. After all, Jaeger could have just as easily called it the 'The Jaeger Convertible Skirt'. First, Alice is described as a 'young and pretty matron'. Given she was able to demonstrate the garment, negotiate business deals and eager journalists and travel vast distances without the need for much rest, she was confident and independent yet clearly feminine enough to pass Victorian muster. Second, the fact that journalists were aware that the design came from 'her own experience' tells us that she was probably a cyclist herself. Using Alice to demonstrate the costume and tell personal stories would have been significantly more compelling to the media and persuasive to pensive women cyclists than if it had been a standard Jaeger product.

In fact, Alice and her invention created such excitement that the story of her business success started to appear in Australian media.[40] One of the more unusual and shortest accounts of her success was in the 'Sporting' column of a Brisbane newspaper, *The Worker* (originally called *The Australian Workman*), a political newspaper affiliated with the Australian Labor Party. The two sentence report starts with a quick summary of the news: 'It is stated that Mrs. Bygrave, who exhibited a cycling skirt at the recent Stanley Show, held in London, went to America with her patent, and sold it for 5000dol.' What is striking about this coverage is that it features amongst boxing, rugby and cycling racing results. This placement goes some way to explain the journalists' clear yet blunt description of how the skirt worked: 'The costume is of the window blind order, that is to say, it is pulled up and down with a cord.' This story, although very brief, provides evidence of the broader impact of Alice's design beyond the London cycling fashion world.

Later that same year the actual 'Bygrave Convertible Cycle Skirt' wound its way to Australia. Jaegar had negotiated deals for its sale in the big eastern cities of Sydney and Melbourne. It features in an advertisement in the *Sydney Morning Herald* on 26 September 1896. The large broadsheet promotion is for David Jones, a large reputable retail store in Australia which at this time called itself 'Specialty Garment Makers for Lady Cyclists.' Here, the Bygrave costume was available in 'Sanitary Tweeds' and cost 29s 6d (this was 8 shillings higher than its UK cost, but probably included a shipping markup). The popularity of the garment in Britain was pushed as a selling point to locals: 'These have only just arrived in the colony, but his patent has been in use in England for some time past, and found much favour.' Although the advertisement clearly uses Alice's name, the advertisement credits her patent to Dr Jaeger.[41]

Alice's skirt clearly appealed to Australian cyclists, as it did to British and American ones. *The Sydney Mail* and *New South Wales Advertiser* described it as an 'ingenious' compromise.[42] 'Life,' lamented the journalist, 'is made up of compromises, but few compromises are so thoroughly satisfactory as this one between the rational costume and the ordinary ladies' attire, which is accomplished by the Bygrave Convertible Skirt.' They felt that this 'ingenious garment' once converted 'allows free pay to the most energetic of lady cyclists.' Interestingly, this writer highlights how the garment can be adapted actually *on* the bike itself: '[A]nd by another single movement, without dismounting it is released and returns to its former condition.' This review also reiterates what has been present in all the media coverage, that a textual and illustrative review did not do the garment justice. To fully appreciate its inventiveness, consumers needed to go and see it in action. Advertisements reinforced the 'indispensible' nature of the garment and encouraged consumers to 'inspect' and have 'explained' the details of its 'ingenious construction'.

> The new invention may be inspected at the office of the wholesaler's agents, Dr. Jaegers Company, at the corner of York and Erskine Streets, where its working is demonstrated on a model. It is worth seeing.

Later that same year, a journalist in *The Australasian*, a Melbourne-based newspaper, was taken by the convertibility of the design. 'Whether riding or dismounted the rider has all the appearance of wearing a tailor-made skirt.' They make a point about how it fits site-specific types of cycling. 'On country roads, the skirt can be modified by means of a few cords so as to give the cyclist the freedom of "rational dress".'[43] Jaegar also took out ads in *The Australasian* to continue to promote Alice's skirt as 'The Latest Novelty' and as 'Indispensible to Cyclists'. It was available for sale in Jaegar stores in Sydney (41 York Street) and Melbourne (314 Flinders Lane).[44]

Was Alice's patent successful? It was clearly popular when measured in press, distribution and commercial terms. According to media accounts it made a significant volume of revenue. But did it work? Was it wearable? From a sociological perspective we can think about what 'works' on many levels.[45] Being endorsed by a substantial textile brand like Jaeger was a remarkable achievement for a first-time inventor. We also see that it was beneficial for the company. Already renowned for its sanitary woolen undergarments and accessories, Jaeger appeared to be using the Bygrave Convertible Skirt to forge a new path into fashionable women's outerwear. By 1897, more than a year after its launch, the success of the Bygrave Skirt was leading reviews of other Jaeger products. Appraising a new saddle cover, a journalist in *The Wheelwoman* writes: 'What with the "Bygrave" skirt, special cycling boots, corsets, etc., this Company is securing quite a large share of the cyclists' custom.'[46]

Did it work on the bike? The many advertisements and editorial encouraged consumers to inspect and see it demonstrated in person. So it must have been convincing in action. The fact that Rosina Lane is pictured wearing it is also a good indication that it did. She was a professional racing cyclist and her livelihood depended on her ability to

JUST OUT !

Novelty for Lady Cyclists !
The Bygrave

" CONVERTIBLE "
CYCLING SKIRT.

Safe, Convenient, Convenante.

All high-class Drapers will supply it.

This indispensable garment may be inspected
and its ingenious construction will be explained
any day at the

Wholesale Agents :

Dr. JAEGER'S COMPANY,

41 York-street, Sydney ;
314 Flinders-lane, Melbourne.

PLEASE CALL.

Figure 6.12 Advert for the Bygrave Convertible Cycling Skirt, *The Sydney Mail and New South Wales Advertiser*, 1896

win races, attract paying audiences and generate media coverage. Her involvement demonstrates not only the functionality of the design but also points to the involvement of Alice's extended family in her work. This might be seen as an early example of commercial sponsorship.

The 1901 Census also provides further evidence that it might have been successful.[47] Records of the night of 31 March document another full Bygrave household. But, unlike the years prior when the boarders were dyers and brushmakers, this time skillsets are predominantly dressmaking. Alice's family of four is still living at 13 Canterbury Road. Charles is 42, Alice 41, Lina 18 and Herbert 15. Alice is (still) not listed as having a job while Charles is reported to be a colonial traveller. There are two servants in the house, Ada E. Fry, 21 (Guildford) and Anna Batram, 22 (Suffolk), and there are two dressmakers, Margaret Paston, 29 (Kent) and Annie Lincoln, 27 (Essex). Was Alice exploring new ideas? Was she making new garments beyond her Jaeger contract? Was she capitalising on her new fame by running sewing classes?

Ten years later, in 1911, Alice's daughter, Lena, has moved out and her son, Herbert, has become an electrician. Both Alice and her husband Charles are listed as boarding house keepers.[48] There are two new servants, Edith Anton, 24 (Peterborough) and Winnie Sims, 21 (Yeovil) and a boarder, Leslie Harris, 41 (Liverpool). Interestingly the boarder's occupation is listed as 'Entertainer – Music Hall Artist'. Victorian society was changing and becoming more open to different forms of livelihoods, whereas prior to 1900 this type of guest would have been scandalous to a reputable household. Perhaps also Alice's and Charles's horizons had broadened with their travel and exposure to the world.

What happened to the money? How much did Alice actually get? Perhaps it was not much. Alice's father, Charles Duerre, died on 20 April 1907 at the age of 77, leaving the total of his will to her alone. This amounted to £366, 18 shillings and thruppence. Although it is a sizable sum, it is nowhere near the money she negotiated for her cycling skirt in the United States. With a total of ten children in the family, it is unusual that the whole inheritance was left to one daughter.[49] This suggests either that Alice was under more financial stress than her siblings or that she needed support to continue her inventive practice.

Another possible explanation for her lack of funds is the prevalence of copies. The 1890s was a highly inventive period in Victorian society and it is possible that similar ideas were emerging and designs were replicated. Owning a patent did not mean the idea was safe. In many cases it was the opposite. Writing about American women cycle wear patent holders, Sally Helveston Gray and Michaela Peteu note similarities between the Hulbert Cycling Suit and the Bygrave Convertible Skirt.[50] The former also featured a series of invisible cords that adjusted the length of the skirt to suit either walking or cycling. The Hulbert was advertised in the highly popular American publication

Godey's Magazine a year after Alice's patent was accepted. And there were others before and after Alice lodged her patents.[51]

Alice's patent for an improved cycling skirt is a remarkable story of engineering, textile design, business acumen and inventiveness. It was shaped by the social context of the time, her personal interests and skills and close connections with family members involved in watch- and clock-making, patenting, dressmaking and professional racing. It is highly likely that her exposure to a wide range of people living in shared accommodation and the happenings in the busy art and design hub of Chelsea also inspired her ideas. Alice imagined, made and patented a device for newly mobile women in the form of a cycling skirt that occupied a 'happy medium' between ordinary dress and cycle wear. It was a more *rational* form of dress but not labelled as such. The patented design provided a means by which women could move between perambulate and cycling identities, and was not just reserved for the aristocracy but as a result of Jaeger's patronage was available to a broader market. It clearly captured people's imaginations. The media surrounding her designs, with accompanying illustrations, editorial, adverts and opinion pieces, along with her regular presence at influential cycling shows, carved out new ways for women to place themselves in public space, not only as cyclists but as successful business women.

Interviewing the 'Bygrave Convertible Skirt'

What more can we learn about this invention by making it? A great deal, it turns out. The patent provides detailed step-by-step instructions for the reader to understand the design. Yet to construct it goes beyond a deep reading of these guidelines to a hands-on, embodied and trial-and-error engagement with the ideas. It is not an easy task. The skirt's internal pulley system is unique and the historical language further compounds the complexity. Directions such as 'leaves the sides of it festooned' initially left the research team perplexed. What kinds of materials and mechanisms produce a festooning effect? These kinds of phrases take on new meanings when you are trying to make them material.

Dressmaking involves the translation of information in multiple dimensions – from ideas with no dimensions, to two-dimensional drawings and block patterns to three-dimensional toiles, which lays the foundation for the final version. It takes time, attention, space and a willingness to re-do things when they don't work. The drawings provided indispensable assistance, moving the maker's gaze from word to image to material and back again. Scaled models and mock-ups were also critical in understanding the nature of the invention. Did Alice make models to develop the idea? She may well have spent long evenings drawing, talking and making models with her dressmaking mother and watchmaking father. Given how well the garment fit her sister-in-law, Rosina was most likely involved as well.

Patents are deliberately structured to make complex knowledge appear ordered, fixed and often flattened. As a result they are compellingly persuasive. Yet, what emerges through making is that this device is not as easily stabilised, and is better considered to be a complex assembly of unstable actors operating in a heterogeneous network.[52] On the surface the Bygrave skirt appears to be an ordinary full-length A-line skirt on a waistband with buttoned side placket for access. The complexity of the garment lies in its engineered infrastructure including two internal pulley systems. While Alice provides three pages of detailed instructions and three technical drawings, the challenge for the research team lay in trying to work out how a successful dress pulley system might operate.

In addition to making the A-line skirt, we identified three main converting elements: stitched channels, weighted hem and system of threaded cords. Following Alice's instructions, two channels were stitched inside the centre front and back of the skirt at the seams through which the cords would run. We initially made these channels from silk lining as per our successful toile. However, we found that wide bias tape worked (and looked) much better. Alice suggests 'tape' or the use of 'suitable guides' and indicates where to locate these devices:

> Both guide and cord are preferably inside the skirt for the sake of appearance; but whether they are inside or outside the skirt does not matter as far as the action of the invention is concerned. It is

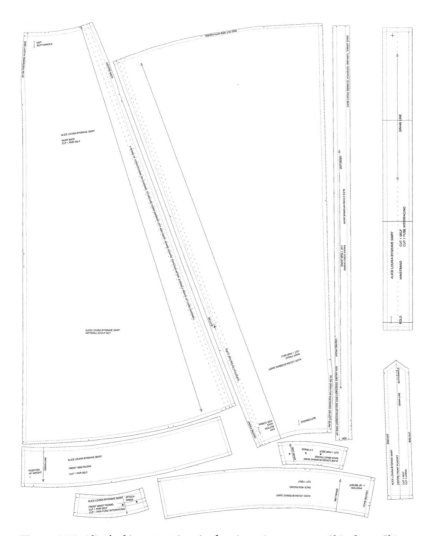

Figure 6.13 Alice's skirt patent inspired a nine-piece pattern: Skirt front, Skirt back, Front cord casing, Centre back cord facing, Front hem facing, Back hem facing, Waistband, Front placket, Front waist facing and Back waist facing

however, of importance that the cord should run and work centrally up and down the front.

The convertible nature of the skirt is achieved with the use of small weights in the hem. Alice is more specific about their purpose and location than about the type or actual weight. She writes how they should be of 'suitable weight' and located 'near each junction of the bottom end of a cord to the skirt.' We chose four round curtain weights with central holes through which each cord could be affixed for extra strength. They needed to be heavy enough to pull the dress back down to the ground but not too cumbersome as to bang against ankles and make walking problematic. They also needed to be sufficiently sewn into the hem so not to fail upon frequent application. This took a few iterations.

Four one-and-a-half-metre woven cords were attached to the weights in the hem and threaded up through the four channels; two at the front and rear of the skirt. At the waistband, the cords fed through six holes concealed under the placket, at the front, sides and inside the skirt. We translated these holes as buttonholes, although no buttons were involved. Alice, again, makes particular reference to the location of the rear cords and buttonholes. They are not located at the back, but rather at the sides for easy access. She explains: 'The presentation of the free ends ... of the cords ... at the side of the wearer, is to make the use of them more easy and graceful than if they both presented behind.' This is particularly clever and highly appreciated when worn, as the wearer is not fumbling for cords dangling at the back of the waistband.

On first read, Alice's suggestions of alternative materials seem to indicate that she is flexible about the nature of elements used. Here are some of her suggestions:

> The guide may be of any suitable type. I have illustrated it as consisting of a hem sewn to the skirt. A row of loops or rings ... Both the guide and cord are, preferably, inside the skirt, for the sake of appearance; but whether they are inside or outside the skirt does not matter as far as the action of the invention is concerned.

> Any type of clip capable of so holding the said cord, may be made use of.

While this can be read as vague, it makes sense when making the garment. What the actual elements are made of is less critical than what they do and the order of mechanisation. This is where Alice's instructions become firm. Her invention was the most complicated garment of the collection due to the interconnected network of elements. Each part

Figure 6.14 Converting Alice's cycling skirt

Table 6.1 Requirements

Skirt materials	Extra garments
2m Dashing Tweeds – Crimson Stripe	Waistcoat
4 x 1.5m woven cord	Blouse
2m printed silk lining	Bloomers/knickerbockers
4m bias tape	Boots, scarf, tights, hat
4 small curtain weights	
4 buttons	

needs to work with the next or the entire piece fails. The cords need to be strong enough to gather two metres of wool material but not too thick as to bulk out the waistband. The guides need to be constructed in clear central lines to enable the ruching of material. The material needs to be the right weight to gather and well-matched to the weighted hem. It would hardly be a 'quick change' skirt if it got stuck halfway. The various buttonholes to channel the cords need to be strong enough not to fray but small enough to remain hidden until required.

Alice is clear about what she wants this convertible dress to do – to give women choice. The skirt was 'proper to wear when either on or off the machine.' Wearers could cycle or conceal their intentions. It is also interesting to note how her invention combines with existing and familiar entities – an ordinary skirt and knickerbockers. She has no desire to re-configure elements that she thinks already work. Her patent focuses entirely with what lay inside. What emerges in the making is how much the skirt operates like a timepiece. There are clear parallels between her invention and her family's watch- and clock-making influences. The seamlessness of the surface belies its complex interconnected and cleverly hidden interior.

Collaboration is a rich theme running through this chapter. Piecing together Alice's story reveals a diverse range of influences that helped shape her creative endeavours; from her watch- and clock-making and cycle racing family to her travels abroad and (local and international) business dealings. Much like her A-line skirt, exploring Alice's life in depth reveals many fascinating layers and valuable contributions to cycling's past.

7

Patent No. 6794: Julia Gill and Her Convertible Cycling Semi-Skirt

Madame Julia Gill submitted a complete specification for a British patent 'A Cycling Costume for Ladies' on 5 January 1895 and it was accepted a little over a month later on 16 February 1895.[1] She lists her occupation as 'Court Dressmaker' and goes by the title of Madame. The patent tells us that she was living at No. 56 Haverstock Hill, in the Parish of St. Pancras N.W. in the county of Middlesex.

Julia's 'entirely novel garment' aimed to 'provide a suitable combination costume for lady cyclists, so that they have a safe riding garment combined with an ordinary walking costume for use when dismounted'. Her design consists of a full-length A-line skirt on a waistband that converts into what she calls a 'semi-skirt'. The mechanism for convertibility in this case is concealed under a second layer or flounce at the bottom of the skirt. Julia suggests using a series of eyelets or rings through which a cord is threaded and loosely tied to keep it out of the way. When required, this corded layer is brought upward from the lower edge of the skirt, gathered and tied at the waist. The result is a bubble-like semi-skirt that sits above the knees. Julia also details two accompanying pieces; a jacket and a bifurcated undergarment in the form of 'fluted trowsers'. Unlike conventional bloomers or knickerbockers, these 'trowsers' were 'tight fitting' and decorated with 'vertical or horizontal flutings or frills'. It is a curious choice, as it seems unlikely that such a design would disguise the wearer's cycling intentions, conceal her independently moving legs or indeed protect her from unwanted social attention.

Some patents, as we have seen with Alice's invention, directly responded to the need for a 'happy medium' between ordinary and more rational cycle wear. This is not the case for all. Patents, as Michaela Peteu and Sally Helevenston Gray have argued, are valuable devices that 'reveal thinking that goes on as a part of the inventive process', however, this does not mean that they 'necessarily reflect reality'.[2] At first glance, designs like Julia's seem at odds with what is known about social convention and concealment tactics at the time (and it seems unlikely that a respectable middle- or upper-class woman would have worn such a garment that revealed this much of her limbs in public). Instead, as I discuss in this chapter, this kind of garment might have been deliberately designed to do the opposite – to stand out.

Cycling in the mid 1890s was increasingly becoming an accepted (and expected) part of a high-class woman's 'accomplishments' that evidenced her social and cultural development. It is possible that a costume like this could been commissioned for the upcoming 'Season' for a young woman to showcase her talents, cultural cachet and ultimately marriage potential. Alternatively, and more likely, given its risqué nature, it might have been a showpiece for the court dressmaker, tactically timed to generate attention. It is entirely possible that concealment was not Julia's aim at all. Instead, showcasing such an exciting cutting-edge convertible invention may have been a distinct strategy deliberately calculated to attract clients and drive new business.

As such, the theme of this chapter is the role of ideas and experimentation in the making of new cycle wear. The Victorian cycling craze catalysed the need for specific garments. Yet, there was no *one* socially agreed or accepted style. New cyclists in need of a costume had a number of options. They could make it themselves by adapting existing garments, buy a commercially made version (like the 'Bygrave Convertible Skirt') or engage the services of a dressmaker or tailor. Julia would have been one of many entrepreneurial business owners in the late nineteenth century offering an exciting range of ideas and designs for women to choose from.

N° 6794   A.D. 1894

Date of Application, 5th Apr., 1894
Complete Specification Left, 5th Jan., 1895—Accepted, 16th Feb., 1895

PROVISIONAL SPECIFICATION.

A Cycling Costume for Ladies.

Madame JULIA GILL Court Dressmaker 56 Haverstock Hill N.W. do hereby declare the nature of this invention to be as follows :—

For improvements in lady's cycling costumes.

The skirt is made with an underlayer of the same material or other kind, which when turned up is drawn in at waist with a cord run through, rings, tapes, or eylet holes &c. which then forms a semi-skirt, the under piece forming a frill & giving the appearance of a jacket bodice. When the wearer gets off cycle the skirt drops into place as an ordinary walking skirt. The same skirt can be fixed up to the waist band or other parts of costume by hooks & eyes or other attachment, or as a combination of skirt, bodice & pants.

The pants are tight fitting in the foundation, with vertical or horizontal flutings or frills, with or without gaiters to finish, thus effectually disguising the form of the limbs.

Dated this 3rd day of April 1894.

JULIA GILL.

Figure 7.1 Excerpt from Julia's cycling costume patent

The Inventor and Her Life

Who was Madame Julia Gill? The initial challenge in studying this patented costume lies with the patentee. Unlike Alice, Julia proves harder to trace through genealogical records. Unless they were extraordinary, of ill-repute or particularly well married in ways that ensured notoriety, there is little written about everyday women's lives in Victorian times. As such, Julia's modest life course is much more in keeping with the reality of ordinary women. Alice is the exception. Julia's case is far more common. This patent therefore presents us with the problem raised in earlier chapters about the difficulty of tracing and keeping track of inventions. Unlike the 'Bygrave Convertible Skirt', no evidence of the 'Madame Gill's Cycling Costume For Ladies' has surfaced (as yet). It may have dwindled in the patent archives or been commercialised and re-branded. It is similarly difficult to trace this inventor's life course. There are so many *Julia Gills* living in London in the late nineteenth century that we

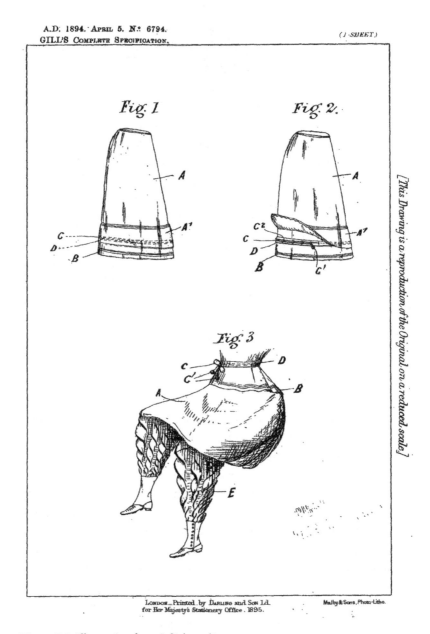

Figure 7.2 Illustration from Julia's cycling costume patent

cannot be completely certain of her biography. Doing historical studies can reveal multiple threads to follow and weave together or overlapping partial pieces and gaps. This is a case of the latter. The challenge is not to write these fragments into a smooth or straightforward story, but rather to resist the notion of tidying up. Cases like this remind us that we can never fully know the past. But, we can try, and see what emerges from various possible scenarios.

One Julia Gill was born in 1848 in Wiltshire. In 1881, she is 32, married to 25-year-old William, a cabman, and living with their three sons and a stepson, in Chelsea. The Census does not list her occupation. Ten years later this Julia is a laundress, her cabman husband and Albert, her youngest, now an errand boy at 14, are all living in Paddington. In 1901, she would have been 53 and the Census tells us that on Sunday 31 March she was a widower living alone in Kensington and making a living doing ironing and washing.

Another Julia Gill was born in 1855 in Clapham Park, London. According to the 1891 census, on Sunday 5 April she was 36 years old, married to Thomas Gill, a solicitor. He was considerably older, at 63. Together they had five children (one of whom died). On the night of the Census, their Kensington house was home to a combination of children and step-children; Messis (32), Thomas (30), also a solicitor, Ada (28), Jessie (26), Beatrice (15), Gertrude (15), Victor (8) and Charles (0). It must have been a significant sized house that required the assistance of a number of employees – a nurse (39) and three servants (47, 19, 16). In 1911, this Julia is 56 years old, widowed, head of the household and living by private means. She resides with two daughters and a son; Beatrice (37), Gertrude (35) and Charles (21). They continue to run a big household with four servants ranging in age from 19 to 30 years old.

Neither of these versions of Julia Gill mentions court dressmaking, nor do their lives fit easily with this vocation. The former is of a lower class and the latter may well have been too busy with such a large family and household to run to operate a separate business. The second Julia, however, does have more links to our patentee. She is older, which matches with the title 'Madame', and her lifestyle fits better with the class of a court dressmaker who could have been working with and for an

Figure 7.3 Our version of Madame Julia Gill's patented 'Cycling Costume for Ladies'

upper-class clientele. Unlike the majority of patentees, Julia did not use the services of a patent agent to process an idea through legal systems. Perhaps the presence of two solicitors in her family provided assistance in this matter. She was also clearly quite wealthy, which implies that she could have been running her own business on the side, and maybe the Census taker failed to list it on the form.

Nevertheless, even without a firm grasp on her life course, Julia's patent, home address and profession provide ample material to explore her invention. Julia self-identifies as a court dressmaker, which locates her in the middle to upper classes. A court dressmaker made high-society clothing worn for special occasions, such as for court events, where young debutants were presented in their finery to the upper echelons of society. Pascoe, in *London of Today*, was somewhat cynical about this rite of passage:[3]

> What is the real purpose of going to Court? Is it, in truth, in order that you may pay your respects to Royalty? Nonsense: in nine cases out of ten, it is in order to see and be seen, to have your name noted by *The Times* or *Morning Post*, to appear more magnificently attired than someone else, to have your dress very fully described in the ladies' journals, to be photographed in that dress at midnight on Bond Street, and generally for the sake of having it known that you have been at Court.

Whatever the motivation, a high-class woman needed to be prepared. She not only needed 'that dress' but several others. Many sought court dressmakers to commission new wardrobes in readiness for the Season. The Season was an annual period spanning Easter through to August where people left their country estates and flocked to major cities, such as London and Bath, to do all kinds of political, social and cultural business. On 1 June 1895, *The London Illustrated* declared that 'the height of the London season' was approaching and 'town is filled with the usual fashionable throng, for whom dances, garden-parties, and receptions are being held in a rapid succession which is dazzling'.[4] No expense was spared and the people who were able to participate in this 'dazzle' constantly sought the newest and most exciting fashions and activities. In *The Party that Lasted 100 Days*, Hillary and Mary Evans argue that the Season acquired cult status 'because leisure society needed this kind of ritual defence against the emptiness of unlimited leisure'.[5] Although the Season has been enjoyed since the regency period in the eighteenth century, it apparently reached 'its zenith in the 1890s'.[6]

Why was it so pivotal in upper-class Victorian life? The Season was a time to 'see and be seen'. Deals were made, contracts signed, debutantes 'came out', 'good' marriages arranged, social connections made and pleasures sought. It was a chance to enter or affirm your place in society and clothing played a primary role. Preparing for the Season was expensive. Power and status was performed and displayed through clothing, and as each planned event necessitated a new garment, every year required a new expanded wardrobe. This was a significant cost and one not shared across society. There an immense imbalance in gendered investment; while men's wear could be had for £10, women's dresses could cost up to £50.[7] (In this context, the Bygrave Convertible Skirt at just over £1 seems very affordable). In *The Art of Dress: Clothes and Society*, Jane Ashelford writes about how the nineteenth century saw men 'universally garbed in discreet, sober clothes in dark muted colours', while women 'had become decorative accessories, proclaiming the family's wealth and status by their display of fashionable dress'.[8] The frequency of Seasonal events sometimes also required several changes of clothes in the same day. Some argued that high society women needed to invest at least £1,000 to meet the exacting standards of the period.[9]

The Season clearly would have made a court dressmaker very busy. Hillary and Mary Evans write about how '[m]any of the tradespeople of Mayfair and the West End streets must have been largely if not wholly dependent on the trade they did during the three months of the Season'.[10] This would account for the timing of Julia's patent during what was traditionally a low winter period. According to her patent, a full specification was completed on 5 January and accepted on 16 February. Late winter was a slack season for dressmakers and tailors. A February column in the weekly trade periodical *The Tailor and Cutter* lamented how '[w]e are now in the midst of what may be considered the dullest season of the whole year'.[11] The author goes on to suggest ways to keep busy; doing a stock-take, chasing overdue accounts and not picking on employees.

Although sport had long been a primary social mediator, via events such as horse racing and golf, at the turn of the century women started to move from spectators to participants. Season events followed suit, broadening women's social activities beyond balls, parties and dinners to include physical activities. Hyde Park was a prime public site

Figure 7.4 Cycling was so popular that some periodicals satirised the fashionable mass uptake by young women, *Punch,* 1897

> *Ethel* – 'I hope bicycling will go out of fashion before next season, I *do* hate bicycling so'
>
> *Maud* – 'So do I! But one must, you know!'

for social activities. Cycling was the 'new cult' in the mid 1890s and increasingly adopted by the aristocracy, who were swapping carriage rides for bicycle rides, which gave the latter social legitimacy and culture cachet.

This marked a significant change, not only in the expanded possibility for social relations but also for clothing. As a writer in *The Queen* explains in 1895: 'For years cycling has been relegated to the sterner sex and the lower middle class. Now that fashionable women are adopting it, the question of the most suitable dress becomes an important one.'[12] A year later, *The Lady Cyclist* announced that '[a]lmost all the royalties at home and abroad have now become bicyclists, and there is hardly a Princess of the blood Royal to be found in England or on the Continent, who has not become a devotee of the wheel within the last two years.'[13] Many popular periodicals filled columns with the social happenings and costumes of the upper echelons of society. *The Lady Cyclist* regularly regaled its readers with sartorial stories of cycling ladies and princesses. Lady Warwick was apparently regularly spotted wearing matching clothing to her bicycles. In 1895 she was attired in an all-white cycle costume to match her white machine. In 1896, she updated both with 'a moss-green suit, with a moss-green bicycle to match' and there was talk that 'another bicycle has just been sent down to Warwick, to her order – chocolate brown, this time, with narrow gold lining'.[14]

As the Season now included cycling, cycle wear became assimilated into wardrobe planning. Even if a woman was not so keen on cycling, as noted by some satirical periodicals, she had little choice but to participate in the most popular Seasonal activities. Women needed to be outfitted for such invitations, and cycling, as we have seen, demanded a specific type of costume that did not get the rider's dress tangled up in wheels and pedals. Early cyclists had several options. They could make or adapt costumes themselves, or engage the services of a sympathetic dressmaker or tailor, who was politically aligned to the idea of a veloci-pedienne. Sometimes women had no choice but to bravely undertake the former, when faced with a lack of the latter. This was not universally encouraged. A columnist in the *Rational Dress Gazette* noticed that a 'considerable majority of the rational costumes worn hitherto have been made by the wearers themselves'. While recognising that this was some-times necessary, this writer remained unconvinced that an individual's enthusiasm could counter lack of skill: 'These wearers in many cases have not even been in the habit of making any of their usual dresses, and possess only a very elementary knowledge of dressmaking'.[15] *The Lady Cyclist* was similarly scathing about low-quality costumes and encouraged women to start to thinking about where to get their new costumes made as early as March.

> Many of you will be getting new cycling costumes soon, and I hope none of you will imagine that it is economy to go to an inferior tailor for them. A good tailor's work looks nice to the last, and cut, fit and finish are three essentials for the smart cyclist.[16]

Madame Julia Gill's court dressmaking services were probably one of many small businesses offering services to new cyclists, amongst other

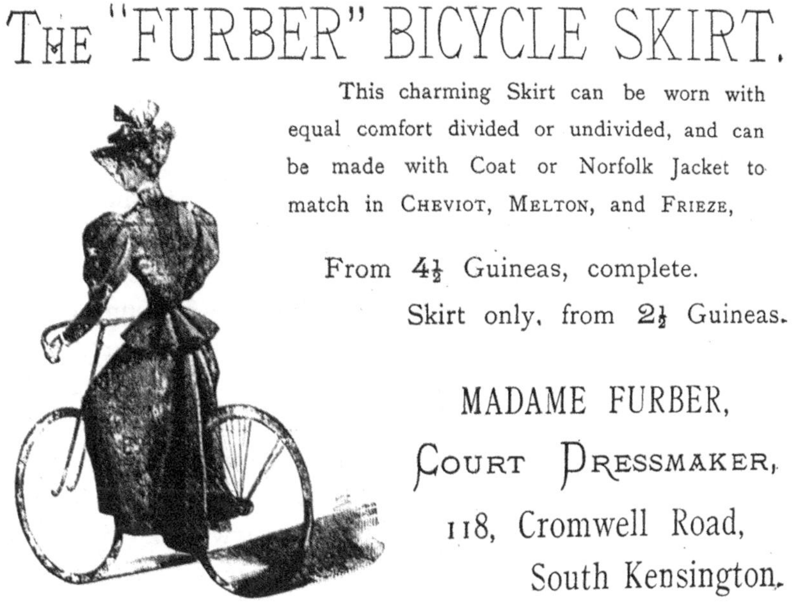

THE "FURBER" BICYCLE SKIRT.
This charming Skirt can be worn with equal comfort divided or undivided, and can be made with Coat or Norfolk Jacket to match in CHEVIOT, MELTON, and FRIEZE,
From 4½ Guineas, complete.
Skirt only, from 2½ Guineas.
MADAME FURBER,
COURT DRESSMAKER,
118, Cromwell Road,
South Kensington.

Figure 7.5 Advert for Madame Furber's patented bicycle skirt, *The Lady Cyclist*, 1896

A.D. 1895. Oct. 13. No 19,191.
FURBER'S Complete Specification.
[This Drawing is a reproduction of the Original on a reduced scale.]

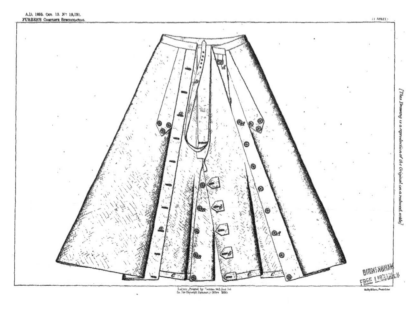

Figure 7.6 Illustration of Madame Furber's patented bicycle skirt

seasonal dress requirements. Through her patented design, we might assume her politics were not far from those of dress reformers, in terms of supporting and enabling women to move more freely in public space. However, she was also a businesswoman, so probably had a keen eye for new ideas. Julia may have been commissioned to make something similar for a client and identified the potential for patenting it. Alternatively, she may have taken the opportunity to pre-empt the pre-Season rush with her own design as a means of attracting a new clientele.

Julia timed her patent well, not only during the slow time of year but also at the start of 1895, which heralded the cycling boom. This was a tactic of other London-based court dressmakers. Madame Evelina Susannah Furber (1858–1945) was also a court dressmaker in the 1890s. She had a shop at 118B Cromwell Road, West Kensington, in London and like Julia she designed a convertible cycling garment for women. She was particularly entrepreneurial. She did not wait to have the patent commericalised by someone else. She did it herself. She produced the garment and promoted it with adverts and editorial in popular cycling and women's magazines. The skirt alone was sold for two and a half guineas or could be purchased complete with a coat or jacket to match for four and a half guineas. This was much more expensive than the 'Bygrave Convertible Skirt'. The price difference is understandable considering the economies of scale that the larger corporation would have been operating within and also because Evelina was offering a tailor-made costume.

The 'Furber Bicycle Skirt' was promoted in *The Lady Cyclist* as a 'charming' and 'novel' skirt with concealed fastenings that created 'a perfect divided skirt if required'.[17] *The Queen* called it 'a most practical invention' because it 'cannot blow up' and assured potential buyers that they would 'have a handsome walking gown as well as a graceful and becoming bicycle dress'.[18] Essentially the 'Furber' was a divided skirt, featuring a seam at the front and back which, when unbuttoned and re-buttoned to an under layer, forms two legs. Evelina explains in her patent how '[e]ach edge of the back opening can also be connected to the same side of the front opening so transforming the skirt into a pair of loose trousers'.[19] A strap, released from the back waistband, passed through the fork to attach at the front and apparently kept these divided legs in place.

The Furber was one of nine new garments from independent dress-makers and tailors showcased in *The Queen*'s 1896 full-page special: 'The New Cycling Costumes of the Season'. This type of editorial reveals the breadth of styles available and the fierce competition of independent businesses vying for reader's attention and buyer's purses. The illustration of Evelina's design shows a woman astride a bicycle, and the pose reveals the subtlety of the transformation.[20] The divided skirt effectively falls over the sides of the wheel, rather than bunching at the front and back. How the wearer might have prevented the flapping material from entering the moving wheel, or exactly what the strap is doing, is less clear. Making and wearing the garment would have revealed much more. However, even at this level of detail, it is interesting to note that despite sharing similar interests and concerns about alleviating the risks of mobile clothed women's bodies, Evelina's garment is vastly different to that patented by Julia. The former displays an understated yet clear desire to maintain and maximise concealment. In contrast, Julia's in 'ordinary' form might conceal the wearer's intention to cycle, but once converted, it was far more theatrical. Contrasting designs like these were really important as they showcased to women a spectrum of ways their differently mobile bodies could be in and move through public space. The availability of these and many other kinds of costumes, must have been remarkably refreshing and also shocking to those more aligned to normative clothing conventions and accepted behaviours. In this way court dressmakers like Evelina and Julia provided critical creative services for women in the 1890s. They equipped them with ideas to imagine other ways of being in the world. They gave them choices.

Ideas and Experimentation

Independent business owners played an important role in the retail landscape in Victorian society (and in helping women negotiate their new mobile identities). Yet, their valuable contributions have been largely overshadowed by radical changes to the commercial sphere at the turn of the century. This was the advent of vast new shopping experiences brought about by department stores, like Selfridges, that consolidated multiple businesses under one roof. They helped to grow the market for ready-made clothing and presented new ways for women to inhabit public space. Erika Rappaport writes in *Shopping for Pleasure* about how London's new West End 'welcomed women into metropolitan culture by configuring public life as a pleasurable commodity and public women as natural consumers'.[21] It is important to consider not only what people were doing at this time but also what kinds of people were doing it. Writing about the history of the beauty business, Kathy Peiss argues that remembering the 'web' of fragmented, small and seemingly inconsequential contributions of small businesses providing bespoke products and services is nevertheless important even amongst these seismic retail shifts. Perhaps not surprisingly, women played a significant role in producing and also challenging feminine ideals at this time and it is the work of 'seamstresses, hairdressers, beauticians, department store buyers, and cosmetic saleswomen'[22] that are least well known.

Court dressmakers like Julia and Evelina would have been typical of these kinds of influential small businesses. They were ambitious, independent women who imagined and made radical new multipurpose convertible garments, and also patented them, and in doing so lay claim to their ideas in larger commercial spheres. Diana Crane writes about how fashionable clothing 'appeared to offer possibilities for a person to enhance his or her social position'.[23] We can imagine this to also be the case for dressmakers. Even if they themselves could not wear high-status objects or be invited to court, they could be associated with them on behalf of clients. Through their work, media coverage and Seasonal associations, they would have developed solid reputations in their industry and made significant contributions to the small business community. They would also have helped to generate 'a female enclave in a largely male commercial world'.[24] In this way, court dressmakers played critical mediating roles in cultural and political spheres. They produced garments that provided an essential service for their clients and also represented a shift from domestic home sewing to public-facing business.

There is also evidence that women's small dressmaking businesses were not just operating on the fringes of larger market forces, but rather actively leading cycle wear innovations. The 1890s cycling craze seemed to take the traditional tailoring sector by surprise. A writer in *Tailor and*

THE "LADIES' TAILOR" CHART OF CYCLING COSTUMES.
Published by The John Williamson Company, Limited. London, W.C.

Figure 7.7 *The Ladies' Tailor* provide options for all political orientations in their 1897 chart of cycling costumes

Cutter reported on the range of cycling garments at the 1895 Stanley Cycling Show and was clearly flummoxed by the shifts they were witnessing. Even at the height of the cycling boom, they found men's garments not very interesting – 'there is not much new to add to what is already known; no new or striking design has been introduced for this particular sport'. Instead, all the exciting invention was to be found in women's cycling clothing where 'great strides have been made' and the market was 'evidently on the boom'. Walking around the displays they saw 'a large number of ladies' tailors, with patent and registered skirts in great abundance' which were 'courting public favour with inventions of a more or less practical nature'.[25]

Two years later, with the cycling craze still in force, the official tailoring sector *still* seemed surprised by the popularity of women's cycle wear. In 1897, *The Ladies Tailor* felt the need to explain its choice to commit an entire volume to cycling in response to the level of enthusiasm from its customers. 'The firm hold cycling has taken on the popular taste during the past few years should be quite sufficient warranty

for our devoting the whole of our Illustrations this month to the newest styles of dress worn for this exercise.'[26] Even more interesting is how
this was achieved. While it identified that there was a 'dress problem', it
refused to directly engage with it:

> Considerable controversy has taken place over the two classes of
> Cycling Costumes more generally worn; the different sides being
> energetically advocated by partisans who generally warmly support
> the particular style they wear to an extent that they desire to exclude
> all other makes from their society ... The battle of the Rational
> v. Skirted Dress will doubtless remain a burning question for some
> time to come, and will only be settled by individual choice.[27]

Instead, it took a more encompassing approach and provided costumes with *and* without skirts. This position revealed their economic
acumen – appealing to both sides of the 'battle'! It worked. A year later,
they noted the 'remarkable development of cycling as an exercise has
given a marked impetus to the Ladies' Tailoring Trade.'[28] They also noted
how the battle of the 'rival styles, viz., Rationals v. Skirts' seemed to be,
in England anyway, leaning in 'general favour of the latter'. Things were
different on the continent, where '[b]loomers are almost as universal as
skirts are in the West End'. The periodical reported that it was 'only the
few intrepid ladies who are prepared to face uncomplimentary remarks'
catalysed by wearing the alternative. Articles like this confirm the deeply
entrenched binary nature of the debate that patentees were questioning
with their designs: did a woman *have* to choose to be a rational or skirted
cyclist? Could she not be both or something in-between?

Regardless of how women sourced a new costume, they needed
ideas to suit not only their politics and social position but also their body
shape, type of cycling, location and of course budget. Ideas and designs
were not in short supply in the late nineteenth century. Businesses,
dress reformers, travellers and the media provided a wealth of inspiration and often contrasting advice. Women drew on ideas circulated via
a range of formal and informal channels. As discussed earlier, newspapers and popular periodicals and cycling magazines teemed with advice
for the new cyclist. *The Queen* featured colour plates of 'The Latest Paris
Fashions' and *The Lady Cyclist* published regular illustrations of 'The

THE IDEAL LADY CYCLIST.
From the point of view of our various Artists.
IV. By George Gatcombe.

Figure 7.8 The Ideal Lady Cyclist, *The Lady Cyclist,* 1896

Ideal Lady Cyclist – From the point of view of our various Artists'. Others provided patterns and competitions for women to win prizes for having the loveliest costume.[29]

Much like today, women would have taken an array of materials to a dressmaker and worked with them to come up with a unique

Figure 7.9 The Latest Paris Fashions, *The Queen*, 1896

customised garment. As Sarah Gordon writes, 'the unfamiliar realm of
sports and sports clothing allowed, even required, a significant degree
of improvisation.'[30] The 1890s was a time of change that enabled women
(including dressmakers) to explore and experiment with the parameters
of what was acceptable to wear for a newly mobile woman. They were

re-inventing forms of mobility from the inside out and using every tool, skill and network they could muster.

New Cycle Wear Retail Experiences

Small businesses showcased not only new and exciting developments in designs to whet the appetites of potential customers but they also competed with each other to entice consumers to part with their money through innovative shopping experiences. Two examples are particularly illustrative: Mr F. J. Vant's new 'En Avant' cycle costume was heavily promoted in *The Lady Cyclist* in 1896. He was a 'Court Tailor and High-Class Ladies' and Gent's Cycling Costumier' located at 67 Chancery Lane, W.C. London. *The Lady Cyclist* was not only taken with his new skirted costume, but also with his newly fitted store – with a dedicated section for lady cyclists. 'A large ladies' show room is being prepared in the basement, with machines on home-trainers for fitting purposes and private dressing rooms.'[31] Again, much like Jaegar's method for selling the 'Bygrave Convertible Skirt', Mr Vant believed that demonstrations were critical. These costumes may have seemed so radically different, that consumers needed not only to see them, but also to feel them on their bodies *and* on bicycles before buying. It proved to be a successful strategy for many cycle wear sellers with try-before-you-buy retailers popping up throughout the central London shopping district:

> The number of really attractive-looking cycle depots which have been springing up in the West-End of late is remarkable. There are now ten where there used to be one, and all seem to be in a most flourishing financial condition, which proves that the surmises which have been afloat recently as to the decrease in the demand for bicycles are apparently unfounded.[32]

A writer in *The Queen* was similarly enthused about Mr Marcus's new retail experience for cyclists which apparently went one step further – it was equipped with an indoor track!

> Mr Marcus, of Conduit-Street, has just opening one of his rooms as a track, where ladies may not only try on their costumes on a machine,

but may make a preliminary trial of their cycles. ... The floor is admirable, being laid with wood upon concrete, and well varnished. As the room is a large size, it is possible for ladies to get a thoroughly good practice on it.[33]

These new in-store experiences may have begun with a practical function in mind but were clearly good business and a lot of fun. To cycle around a store in new forms of clothing on brand new bicycles would have been a thrilling way to shop. It would have offered a chance to meet with friends, to see the newest Season's costumes and velocipedes, and to experience them in action (without fear of harassment). These spaces marked a radical shift in the shopping experience – whereby clothes were beginning to be assessed not only by how they looked, but also by how they enabled or constrained the body while moving.

We have no evidence that Madame Julia Gill's patented cycling skirt was made, commercialised or ever even left the patent office. Given its risqué nature, it seems unlikely that it would have been a popular costume for public use. However, cycling was not the only reason to purchase a cycling costume. There is evidence of cycling costumes being purchased for other purposes. Julia's garment may have been commissioned for fancy dress or private bicycling parties, that ocasionally did not involve riding at all but rather epitomised fashionable consumption of cycling culture in high society. Some women purchased costumes to have their bicycle portraits taken as a personal keepsakes and as a more controlled form of participation in this exciting new activity (more about this in Chapter 9), or to cycle in private gardens or indoor cycling schools. It is also possible that Julia could have made and showcased it in her home or shop as an exemplar of her skill. As such, it might have appeared in a public-facing window, a primary place of promotion for small business and a growing platform for the developing consumer culture in Victorian society.

The Great Exhibition of 1862 had generated an appetite for walking and visually consuming things. The sensory delights of West End London shop windows apparently produced the 'city as exhibition' for people to enjoy on a daily basis.[34] The pleasure of this urban experience is captured by a Victorian journalist, who explains: 'There are few people

Figure 7.10 Mr Marcus's New Cycling Show Room, *The Queen*, 1896

who have not been struck with the magnificence of the London shop-fronts. They form one of the most prominent indications of the grandeur and wealth of the metropolis.'[35] The persuasive power of window shopping was not lost on British tailors. *The Tailor and Cutter* often discussed the critical role of retail windows to attract passing trade: 'The shop window is, as everyone knows, dressed well and carefully, as an advertisement for the establishment ... an indication both as to price, style, quality and other matters of what is to be found inside.'[36] These spaces were not just about showcasing new clothing, but about inspiring new imaginings and staging futures. Small business owners like Julia may well have used their shop windows to furnish women with possibilities for thinking about new ways of being in and moving through public space.

Interviewing Julia's Convertible Cycling Semi-Skirt

Julia's patent is for an ordinary full length A-line skirt that converts into a semi-skirt. It has a short opening on the side, with buttons to close, a narrow hem and waistband. There are no darts, though they can be added to assist with the fit. The main detail lies in the eight-inch flounce that overlaps the skirt's lower edge. Julia explains how the wearer can 'run a cord through rings eyelet-holes, tapes or tuck with the bottom flounce'. The garment is converted by 'turning up the skirt and attaching it at the waist by cords, or rings, or other attachment'.

The patent provides ample instruction, yet even with the detailed drawing, it was the least clear of the collection. Julia's explanation only made sense to the research team when we made, put the garment on and attempted to convert it. We chose to use her first suggestion of cord and rings. The convertible device – a one-and-a-half-metre cord and 15 curtain rings – is hidden between these two layers at the skirt edge, and remains undetected until activated. To convert the skirt into cycle wear, the wearer lifts the lower edge to the waist, tucking in the flounce inside the folded area, and gathering the cord to create a semi-skirt (it helped that some of the research team remembered 1980s bubble skirts).

Although Julia mentions three pieces in her patent – jacket, skirt and 'fluted trowsers' – the patent is only for the convertible skirt. Yet, she has clearly thought about and provides direction about how the skirt relates to the other pieces. This relationship becomes all the more evident when all three are constructed and worn as an ensemble. Julia's dressmaking sensibilities also surface in the attention she gives to choice of materials, which are simultaneously revealed and concealed in the process of conversion. 'The skirt is made with an underlayer of

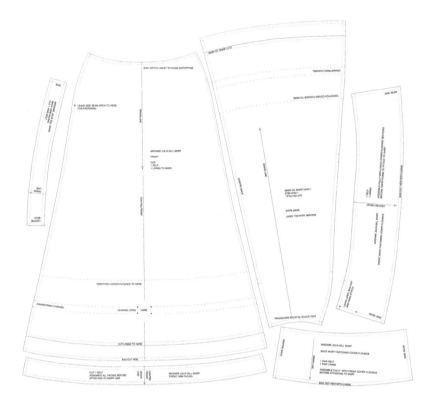

Figure 7.11 Julia's patented skirt consists of six pattern pieces; main skirt body, two-part hem and collar facing. We added a small rolled waistband and a buttoned side placket to allow for entry

the same material' and 'the under piece forming a frill and giving the appearance of a jacket bodice'. In theory, this was not so clear. But in practice, it is very clever. By specifying that the hidden layer is the same pattern as the wearer's jacket, when gathered at the waist, the main skirt material disappears and the flounce becomes a matching frill or second peplum to the jacket.

Julia also specifies details of the bifurcated undergarment. She describes it as 'tight fitting in the foundation, with vertical or horizontal fluting or frills'. We were less convinced by this detail, in theory and in practice. We studied the description and illustration closely, as there were no other obvious references to such a garment in popular periodicals of the time. There are many ways to interpret this and our combination of conventional tights and twisted ribbon detail gets somewhat close to Julia's description. Julia suggests that this design operates by 'effectively disguising the form of the limbs'. Making and wearing this garment suggests otherwise.

The skirt is unique in that it dramatically changes from an ordinary floor-length A-line skirt into an above-knee voluminous bubble-shaped skirt (the patent illustration does not quite render the real height of the fold – it is higher). While the former is unobtrusive, the latter could not be more obvious. The bubble shape draws attention to the waist and the legs are greatly exposed. At a time when the mere glimpse of a woman's independently moving legs was unseemly, this would have been risky and very experimental. Like Alice's skirt, Julia's invention has a quick-release mechanism to enable the wearer to assume her 'ordinary' walking appearance 'by simply unfastening the cord c and dropping the turned up portion of the skirt A'. This works very effectively (which is just as well considering the exposed limbs). Thinking like a dressmaker, Julia's design also deliberately minimises the potential for telltale creases. She explains: 'The arrangement of tuck D and flounce A presenting any creases appearing in the skirt when let down for walking' (it is worth noting that no other inventor discusses the problem of creasing).

More than any other in this collection, this skirt really did not work off the body. We were initially perplexed by the drawings. Then, we were underwhelmed when we first looked at the completed toile. It was only when we took turns to try it on and convert it, and then cycled in it that the radical transformative potential of the skirt really became evident. This is the only skirt that completely inverts. The inside becomes the outside; a process that has an unexpected advantage. Practically, while cycling, the inside of the garment would catch dust and mud splatters from the road. When the cyclist transformed the garment back to street wear, all of these marks would be effectively hidden from view. This insight emerged during one of the public 'Show and Tell and Try on' events run during the project. After telling stories about Victorian inventors and their cycle wear patents, we would invite people into the garments. On one such occasion, a woman tried on Julia's skirt, transformed it a few

Figure 7.12 Converting Julia's cycling semi-skirt

Table 7.1 Requirements

Skirt materials	Extra garments
2.5m Dashing Tweeds – Sea Green Raver Wave	Jacket – Dashing Tweeds, Shetland Jig
1m Dashing Tweeds – Shetland Jig for flounce	Blouse – green and white stripe cotton
1.5m green ribbon or cord	Fluted 'trowsers'
15 curtain rings	Boots, scarf, tights, hat
2.5m printed silk lining	
4 buttons for side opening	

times and realised this advantage in relation to the mud and grease on her own trouser leg from the cycle ride to the event.

Julia calls her design a 'perfectly, safe, easy and graceful riding garment'. We agree in that the height of the folded skirt greatly reduces the chance of material catching in the moving rear wheel. The hem is gathered up and folded at the waist: it cannot blow up or out. The flounce effectively hides the gathered cord threaded through the concealed series of rings when in the lower position and also when tied at the waist. It also transforms into a double peplum with the jacket, the effect of which is visually pleasing. And while it takes some practice to convert it into a cycling semi-skirt, it is comfortable to ride in and converts back to the walking skirt very quickly. However it is also, in comparison to other patents, uncompromising in its experimental nature. On paper this garment looks socially possible, but in material, it reveals itself as very risky!

Even without firm evidence that patented designs like Julia's were made, sold or worn, they nevertheless remain important sources of data about Victorian women's inventiveness. In particular, Julia's story provides valuable glimpses into the lives of court dressmakers – what they were doing and thinking about over a century ago, and how their actions materially intervened in discussions about women's engagement and rights in public space, both for themselves as business owners and for independent mobile women.

8

Patent No. 8766: Frances Henrietta Müller and Her Three-Piece Convertible Cycling Suit

Frances Henrietta Müller (who went by Henrietta) submitted a complete specification for 'Improvements in Ladies' Garments for Cycling and Other Purposes' on 25 April 1896 and it was accepted on 30 May 1896.[1] She self-identifies as a gentlewoman, which was broadly defined in the late nineteenth century as a woman of high social position, and was living at the time in Meads, Maidenhead, in the county of Berkshire.

What makes this patent unique in this collection is its attention to an entire cycling costume. Henrietta considers not only the skirt but also how it interacts with outer and undergarments and, unlike Julia, she patents all *three* garments – a fitted knee-length coat, an A-line skirt that converts into a raised garment via loops and buttons at the waistband, and a vest and knickerbocker combination. The three pieces work together to enable a woman to look 'ordinary' and respectable in polite society and also to convert when needed into a safe and comfortable cycling costume. She explains: 'The whole suit forms a knickerbocker costume with all its convenience, yet which may be wholly or partially disguised at the will of the wearer, and admits of freedom in riding a diamond frame machine if desired, with facility for a return to more ordinary costume if wished at resting places, by releasing the looped skirt.' While it is one of the more conservative-looking assemblies, this 'disguise', to use Henrietta's words, all the more cleverly conceals its transformative potential.

The theme that emerges from archival research into Henrietta's life relates to women's fight for emancipation and how this shaped her understandings of clothing. As will become clear, Henrietta dedicated her life to women's suffrage and promoted freedom of movement in many aspects of social and political life. She was physically active and enjoyed dancing and riding when she was younger and mountain climbing, rowing and cycling as an adult. Given a woman on a bicycle in public swiftly became shorthand for the 'New Woman' who agitated for change, it is easy to see why Henrietta applied her political beliefs to cycle wear. Although she is the most well-known woman in this collection, as a result of her suffrage activities in Britain and America, her convertible cycle wear patent has not, until now, been linked to her other considerable achievements.

The Inventor and Her Life

Exploring Henrietta's life course provides a valuable framework upon which to locate her patent and gain a sense as to how and why she chose to design such a seemingly complex three-piece cycling suit. Henrietta was born in Valparaiso, Chile, in 1848 and had a rich, diverse and multicultural upbringing that exposed her to ideas and people from around the world. In addition to the patent and media accounts of her many activities, this chapter benefits from writings by Henrietta herself. Much of this first-person account comes from the *Woman's Herald: Women's Penny Paper* published from 1888 to 1893, which she founded and edited. We get to hear in her own voice some of the things that she felt important or frustrating, and usually both, and how she responded.

Henrietta's mother was Maria Henrietta (née Burdon) who was also born in South America but of English descent. Her father, William Müller, was from Gotha in central Germany. Both were incredibly influential in her life. Her mother was committed to the suffrage movement, as a member of several high-profile associations such as the Society for Promoting Employment for Women and the Central National Society for Women's Suffrage. Henrietta thought highly of her, and wrote that she 'sympathises with me very greatly in my efforts and has always done what she could to encourage a spirit of freedom and independence in her daughters; she

N° 8766. A.D. 1896.

Date of Application, 25th Apr., 1896—Accepted, 30th May, 1896.

COMPLETE SPECIFICATION.

Improvements in Ladies' Garments for Cycling and other Purposes.

I, FRANCES HENRIETTA MÜLLER, of Meads, Maidenhead, in the County of Berks, Gentlewoman, do hereby declare the nature of this invention and in what manner the same is to be performed, to be particularly described and ascertained in and by the following statement :—

5 These improvements consist in the form and combination of three specially constructed articles of ladies' costume, so made as to afford special facility and convenience when cycling.

Reference is made to the accompanying drawings in which

Figs. 1 and 2 illustrate a combination vest and knickerbockers.

10 Figs. 3, 4, 5 and 6 refer to a skirt, and Figs. 7, 8, 9 and 10 show an outside garment or coat.

Referring to Figs. 1 and 2 which form respectively front and back views of a combined vest and knickerbockers, it will be seen that the fastening is made by buttons A down the left side, and that at the back a fold B may be let down

15 by undoing the buttons C. The knickerbockers are long enough to extend below the knee and may be provided with strap and buckle, buttons, or elastic band, to retain them at this point. They may be of more or less fulness according to the taste of the wearer.

Figure 8.1 Excerpt from Henrietta's three-piece convertible cycling suit patent

is a woman of remarkable originality of character.'[2] Henrietta's father was a successful businessman in Chile and continued to do well in Britain. An 1871 business register records William Müller, Esq of Hillside, Herts as a Magistrate for Middlesex, the Liberty of St. Albans and a commissioner of Income Tax. As was the Victorian convention, it lists him as having 'an only son, William.'[3] While this is true, the record omits his other children. He had three daughters as well. Henrietta had an older sister, Wilhelmina, and a younger one, Eveline.

Henrietta's early travels laid a foundation for a peripatetic life. 'Both my father and mother', she recounts, 'were very fond of travelling and I cannot count the number of times I have been on the continent.'[4] The Müller family left Chile when Henrietta was nine, and travelled to Boston, and then onto London, where they lived for two years. They moved back to Chile briefly before returning to London where they then stayed. During this time, the children's initial education was informal but nonetheless important. On 7 April, according to the 1861 Census, the family had a governess living with them, Fanny E. Burton, 47. All the

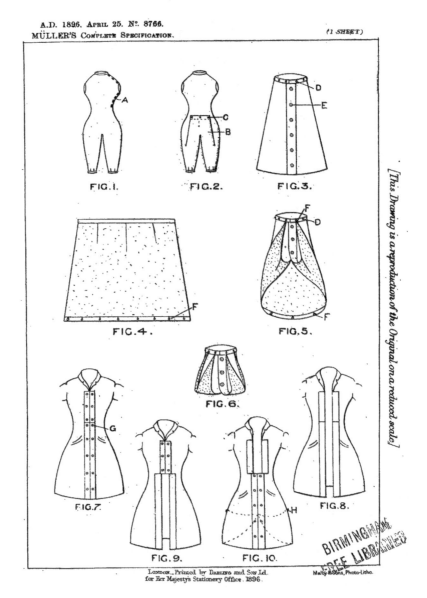

Figure 8.2 Illustration from Henrietta's three-piece convertible cycling suit patent

children are listed as 'scholars', even Eveline, who was eight at the time. Henrietta notes how she enjoyed her education: 'I was rather quick at my lessons, and therefore my teachers liked me and were extremely kind to me.' Her education must have been thorough and wide-ranging, as she spoke six languages, French, Spanish, German, Italian as well as some Latin and Greek. Henrietta's training provided a solid grounding for her ongoing studies and academic life. In 1873, at the age of 27, she gained entry to Girton College to study Moral Sciences, which include Political Economy, Philosophy and Psychology. She stayed there for three years.

Only a few years earlier, in 1869, Girton College in Cambridge had become the first establishment to offer education to women to degree level (Newnham College was opened shortly afterwards). Girton was the first to allow women to take the Tripos Examination that normally enabled a (male) student to qualify for an undergraduate degree. Henrietta was one of many women students to undertake and pass this exam. Yet they were not allowed to gain equal degrees until well into the next century, despite matching and in some cases even beating their male peers. In 1890 Philippa Fawcett, the daughter of a leading campaigner for women's rights, Millicent Fawcett, was classed *above* the top male student in the mathematics Tripos at Newnham College. In response 'to the question as to whether she thought there was anything in the cry that we were educating the mind at the expense of the body, Mrs Fawcett said, "No, I think it's all rubbish" '.[5]

Women's equal access to education was impeded by the same fears that surrounded women's cycling: the concern that it could potentially distract or even worse impair a girl's ability to perform her primary moral duty, that of securing a husband and raising a family. Fears of an education damaging a woman's health were debated at length. Many doctors and psychiatrists believed that women's desire for independence could lead to sickness, loss of fertility and even death. Henry Maudsley, a much-referenced Darwinian psychiatrist of the time, was particular vocal about the immense injury women could suffer if they challenged the 'natural order'. In his 1874 essay *Sex in Mind and Education* he argued that sex was in the brain and linked to reproductive organs, and this inhibited women from achieving equal lives to men. He was adamant that '[i]t would seem plain that women are marked out by Nature for very different offices in life from those of men, and that the healthy performance of her special functions renders it improbable she will succeed, and unwise for her to persevere, in running over the same course at the same pace with him.'[6] He argued that women should not attempt to undertake the same mental tasks as men or else suffer immeasurable pain and untreatable injury. According to men like Maudsley, the ramifications of allowing women to gain a proper education were grave. It endangered not only a woman's health but also undermined social cohesion, and potentially threatened the entire human race.

> For, it would be an ill thing, if it should so happen that we got the advantages of a quantity of female intellectual work at the price of a puny, enfeebled, and sickly race. In this relation, it must be allowed that women do not and cannot stand on the same level as men.[7]

Elaine Showalter's vivid account of *Women, Madness and English Culture from 1830*, as discussed earlier, explores in detail the perceived threats to social order catalysed by women's desire to break free from repressive patriarchal systems. She explains some of the horrific consequences many believed would unfold if women enacted their desires for change: 'Mental breakdown, then, would come when women defied their "nature", attempted to compete with men instead of serving them, or sought alternatives or even additions to their maternal functions.'[8] These social fears, as voiced by experts like Maudsley, were hard to shake off. Many believed that 'once it appeared, mental disorder might be passed on to the next female generation, endangering future mothers.'[9]

Women's education institutions were at the forefront of these debates. They were providing the very thing that many felt would doom society. To protect themselves from unwarranted social attacks, these establishments had to ensure there was no cause for more attention than necessary. This meant that women students' lives were much more strictly monitored and controlled than those of their male peers. In a study about *Women at Cambridge*, Rita McWilliams-Tullberg writes about this impact on female students, at a time when they should have been revelling in the freedoms of this new opportunity: 'The lives of the

Figure 8.3 Our version of Henrietta's converted suit in action

women students were ordered by innumerable small rules of behavior, and it is only possible to understand how the students bore with these constraints by considering the narrow lives they would have had to live had they stayed at home.'[10] Even Henrietta, who came from a family seemingly more open to travel, ideas and women's freedoms, had to fight to gain access to further education. 'After a great deal of difficulty and opposition from my family', she writes, 'I managed to go to Girton where I spent three most happy years'. However, like her fellow students, she too found life on campus stultifying, such that 'the tone of the place is narrow and there is a great want of outside social life to act as a relief from the intense hard work of the students'.[11] Nevertheless Henrietta's Girton years were deeply formative in developing her interests in women's suffrage. She was a passionate and prominent women's rights activist and devoted her life to the advancement of women's freedom of movement in all spheres. The hanging of the bloomer-clad female cycling effigy by male students in protest to women gaining full degrees at Cambridge University took place a decade after Henrietta finished her schooling. The continued hostile reaction to women's desire for an education no doubt would have only served only to fuel Henrietta's motivation and political actions that she believed were instilled in her from birth.

> I really cannot say how I came to feel as I do on the subject of the emancipation of women. I think it must have been born in me. During my girlhood I saw and felt a good deal of masculine tyranny ... I made the deliberate choice to devote myself, body, soul and spirit, to what was then the unpopular cause of women's emancipation. I have never swerved a single instant from that position.

Henrietta was a woman who put her words into action. In 1876 she launched a women's printing press with two others, Emma Paterson and Emily Faithful. The printing press hired 40 women to print materials for women's suffrage and paid them equal rates to men doing similar jobs. She was also committed to campaigning for women's right to vote, for which she was arrested in 1884. Henrietta was close to her sister, Eveline, or Eva, and enjoyed similar interests, in regards to feminist activities, sport and travel. They lived together in Cadogan Place, in West London, which gained much media attention when they refused to pay council tax on the grounds that they, as women, were being taxed without representation. They were arrested and their goods taken by bailiffs and sold. Henrietta explains her actions in a letter to the editor of *The London Times*:

> It is not necessary for me to enter into any arguments upon the merits of woman suffrage. All that I desire to do is to offer a few remarks

in my own justification ... I should like to ask those who disapprove of my action, what course was open to me compatible with my conscience. The principle that REPRESENTATION ACCOMPANIES TAXATION has been the basis of every argument used by me during the last six or seven years when pressing the claims of women to representation. It appears to be urged by my critics that on the first application of a practical test to the sincerity of my words I should have abandoned my position and thereby admitted the feebleness of my convictions.[12]

Henrietta continued to agitate for women's emancipation on a number of platforms. She campaigned for education reform through the London School Board for 12 years from 1879, against sexual violence towards women and girls, for equal pay for equal work in 1883, and even argued for contraception to free women from continual child-bearing in 1884. Although she was a regular writer for *The Westminster Review*, she became so frustrated by the lack of women's voices in the media that she felt she had to do something more substantial.

One of the things which always humiliated me very much was the way in which women's interests and opinions were systematically excluded from the World's Press. I was mortified too, that our cause should be represented by a little monthly leaflet, not worthy of the name of a newspaper called the Women's Suffrage Journal. I realised of what vital importance it was that women should have a newspaper of their own through which to voice their thoughts, and I formed the daring resolve that if no one else better fitted for the work would come forward, I would try and do it myself.

Henrietta set up a weekly periodical, the *Woman's Herald*, in 1888 (later called *The Woman's Signal*). She edited it for five years under the name Helena B. Temple from a central London office at 86 Strand W.C. She explained the reason for the pseudonym 'was in order that my own individuality should not give a colouring to the paper, but that it should be as far as possible, impersonally conducted and therefore open to reflect the opinions of women on any and all subjects'. The newspaper covered a spectrum of issues and subjects related to the suffrage cause and women's rights such as law court news and legal challenges, maternity

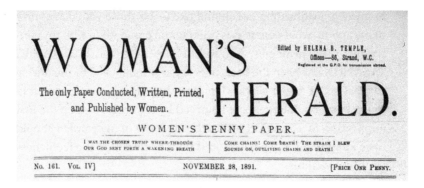

Figure 8.4 Women's newspaper launched by Henrietta from 1888

rulings and parliamentary debates and types of women's employment. It also featured interviews with hundreds of famous emancipists such as Millicent Fawcett, Frances Willard (President of the National Women's Christian Temperance Union) and Elizabeth Cady Stanton (President of the American Women's Suffrage Association). The newspaper banner defiantly declared it to be: 'The Only Paper Conducted, Written, Printed and Published by Women.' It also featured articles on dress reform, patenting and cycling. In fact, the very first edition, on 27 October 1888, featured a column discussing the popularity of cycling for women despite its critics.

> Cycling has become very popular with women and it is a rare thing to hear of any one being injured in healthy by it, though there have been, and still are, plenty of people who cry it down as the unwomanly exercise calculated to inflict injury. Curiously enough the people who utter these sentiments are generally those only fit, as someone expressed it, 'to be dragged about by horses'. As a matter of fact cycling has a most invigorating effect on the constitution.[13]

The second edition in November included a 'Current News About Women' column that became a regular feature in subsequent weeks. This first piece included news of a recent patent by an American woman, Mrs Emma D. Mills, who invented a typewriter attachment and was planning to manufacture and sell it herself.[14] The *Woman's Herald* continued to dedicate column inches to many issues and services relevant to women's dress, such as accounts of Rational Dress Society meetings discussing the 'disadvantages of the present system of dress',[15] adverts for 'Divided skirts ready made or to order',[16] as well as the services of local dressmakers and milliners.

Henrietta continued the nomadic life she was born into by travelling extensively as an adult. She visited nearly every country in Europe, as well as being a frequent visitor to India and America, and regularly gave lectures about her travels and women's rights. She became involved in the Theosophical Society and discussed her theories about women's freedoms in relation to religious doctrines all over the world. This developed her reputation and, according to this report, impacted on her health:

> Miss. F. Henrietta Müller, owing to prolonged ill health her active, hardworking days are over for the present, though her interest in all matters, political and social are as keen as ever. Only a few weeks ago I visited her in the charming cottage near Maidenhead, Berks, in the quiet of which she hopes to regain her strength. It is a real old English red brick cottage standing back from the lawn in a delicious over-grown garden on the borders of a great common, and furnished within with a simplicity equally appropriate and artistic. Here, far away from the rush and roar of London, Miss Muller reads and writes and sees a few friends. She talked eagerly of her past work and on the whole subject of the independence of women, of which she is such a determined and able champion.[17]

In 1891, she was 44 years old and still residing in her 'charming cottage' in Pinkneys Green, Berkshire. The Census records her vocation as 'Editor' and she was 'living on her own means'. She clearly overcame this reported period of ill-health as she continued to travel. In 1895, she is reported to have returned from a third visit to India with a Hindu boy, Akrhaya Kumara Gosha-Müller, who she had adopted. He went on to study Law at Cambridge, and she left London again in 1901 and travelled around the world, on her own. She was still giving public talks at the turn of the century.[18] One of these talks was advertised for women only, and invited guests included Susan B. Anthony as well as State Presidents and Delegates of the Members of the National American Women's Suffrage

Association. The topic was *Free Motherhood or Parthenogenesis* (which is a form of female-only reproduction without the need for male genetics). In these events she put forward her views on 'the steps necessary to the final realization of your labors and to the immediate attainment of complete freedom by the women of America, and through them, of the womanhood of the world.'[19] By all accounts, Henrietta was a force to be reckoned with. Elizabeth Cady Stanton, the famous American social activist and women's right's activist, described her as 'fearless, aggressive and self-centred' and 'she has a luxurious establishment of her own, is fully occupied in politics and reform, and though she lives by herself she entertains her friends generously, and does whatever it seems good to her to do.'[20]

Henrietta was 50 years old when she registered her cycle wear patent. She'd already lived an incredibly productive life when she directed her attentions towards women's clothed mobile bodies. Given the breadth of her interests it is easy to see why she focused not just on a single item, but on an entire three-piece interconnected costume. She was committed to the idea of progress for women, and not content with trying to fix one element, when she could see problems with the entire system. Few technologies at this time could match the potential freedom represented by the bicycle. Attending to clothing was one way of directly addressing the barriers facing newly mobile women and encouraging them to embrace this new freedom of movement.

Unlike the other women in this collection, Henrietta's life is the most explored and readily available online and in archives. Her contributions to the feminist cause are significant. Yet, still, given the breadth of her activities and coverage, her name remains largely unknown. There are no statues or plaques to commemorate her tireless work for the emancipation of women. At the time of writing Henrietta's patent had not been linked to her suffrage identity. The Cambridge Orlando Project has collated the most substantial collection of her many works, and researchers hypothesise why this is the case:

> HM [Henrietta Müller] is less known than one would expect, given her extensive activism in the early feminist movement and her importance as the founder of the first English newspaper produced

by and for women. The fact that most of her work was produced as journalism, and did not appear in volume form, is likely one reason that her trenchant analyses of the situation of women in the later Victorian period have not received greater attention.[21]

Henrietta died of pneumonia at 11 a.m. on 4 January 1906 at her Washington home. She was 59 years old. She still kept a London address in Westminster. Her sizable estate, valued at £12,750 15s 7d, was left entirely to her sister, Eva. The death notice lists her as a 'Middlesex Spinster', which is a depressingly limited descriptor for someone who led a remarkably accomplished, brave and dedicated life in support of women's rights.

The Gender Politics of Pockets

Henrietta's commitment to women's freedom of movement is materialised not only through the design of a convertible cycle costume but even more specifically through her suggestion that the wearer should include as many pockets as was desired. She writes: 'As regards pockets, I find it a convenient arrangement to put two watch pockets, one on each side, in the breast of the vest, a single pocket in the skirt, and two in the coat, though extra pockets according to special taste may of course be added.' Henrietta's cycling suit has at least five sewn pockets, and even more become evident when wearing and converting the garment. Some were for display. Some were practical, while others were designed and located for more private, hidden purposes. The volume and mix of these is interesting when considering the history of pockets.

Pockets are political and gendered objects. Although small, mundane and easily overlooked, they play an important role in the construction of mobile bodies and gender relations. Pockets, and moreover their absence (even in contemporary women's fashion), have long held powerful practical and symbolic value. They provide self-sufficiency and security. They hold objects which free hands to do other things. As Barbara Burman and Seth Denbo argue: 'The things we carry with us on a daily basis reveal a lot about the pace and complexity of our lives.'[22]

They point to roles and responsibilities, of capabilities and capacity for social, cultural, financial and political action.

Pockets for men have been sewn into garments since the seventeenth century. They were rarely hidden as they were displays of power and property. For women, pockets were often separate to their main garments, constructed with ties that could be worn under skirts and moved between garments. They were specially sewn, handed down, gifted, purchased and also, because they were separate objects, easily stolen. Pockets were made and worn by all classes. They could be inherently ordinary and worn much like underwear. They could also be heirlooms, part of ritualistic practice and gift exchange, passed on from generation to generation. For women, a pocket was a prized personal and private space. For many centuries people, and particularly women, lived very closely to one another, often with little personal space. For women who have traditionally owned few personal goods, and were legally not able to own property and were more often considered property themselves, pockets were a way of keeping something private. Ariene Fennetaux, who writes about women's pockets in the eighteenth century, argues that pockets were 'one of the few places women could call their own'.[23] Therefore, the changing nature of pockets, in particular their move from inside to outside garments, and in Henrietta's case, to the desire for an increasing volume of them in women's clothes, 'testify to their increasing mobility and independence'.[24]

Pockets became critical in the context of active lifestyles in the late nineteenth century and with the introduction of new mobility technologies, such as bicycles. Women were starting to move more and their hands were required to manage machinery. As such, pockets started to take on new shapes and different kinds of use. However, as Barbara Burman writes, '[t]here is evidence from the extant clothing itself, and elsewhere, that women were not able to rely on practical pockets being readily available in their clothing, whereas a man could assume his ready-made or bespoke suit or coat would be liberally provided with pockets'. This meant that women became more creative in regards to the nature and location of pockets, such as 'concealed pockets sewn inside hand muffs, travelling rugs and foot muffs for use in trains or later in motor cars, and travelling bags and hand-held cases designed with pockets and straps to secure small personal possessions'.[25]

One of the more unusual responses to the practical pocket problem was in 1885, when Madame Brownjohn of 35 Churton Street, Belgravia, designed a cycle costume for ladies, which was so highly regarded that it was displayed at the International Inventions Exhibition. It was part of a World Fair held in South Kensington, London, featuring cutting-edge design and industry from Britain, America, Italy and Japan that attracted four million visitors in six months. Madam Brownjohn's design, the *Cripper Ladies Tricycling* costume, was awarded a medal for its ingenuity. One of its lauded features was the use of pockets: 'A number of pockets are placed in the dress, one of which (placed inside the under-skirt) is sufficiently large to carry all the luggage a lady might require if she intended to stay the night at the end of her journey.'[26]

We can only imagine what a pocket large enough to carry a lady's luggage might have looked like, or how it might have felt to cycle while wearing such a pocket, but it points very clearly to how women were creatively re-configuring materials, technologies and their bodies to carve out new forms of independent mobility. A decade later, pockets were still as important for mobile women. Those making their own cycle wear were often encouraged to add useful pockets. Marguerite from *Bicycling News* offers this advice about pocket positioning for the budding dressmaker:

> The usual place for a pocket is in the back of a skirt, but this is most uncomfortable for riding. I have found it a good plan to have two pockets – one let into the skirt in the ordinary way, but not quite so far back as is customary; the second, an outside pocket in which the handkerchief can be carried and can be got at without any trouble.[27]

Pockets were also of interest to Victorian inventors. The turn of the century saw an increase in patents for pockets that reflected a change in society; people were increasingly moving outside the home seeking employment, combined with the increasing proximity in which people lived and socialised as a result of the industrial revolution. With pickpocketing rife in city centres, inventors were motivated to find ways to protect people's personal possessions. Inventions of this time include Amy Hart's 1895 patent for 'Improvements in Pocket Protectors' and

Figure 8.5 Postcard of a model wearing Madame Brownjohn's tricycling costume with hidden pockets, 1885

Blanche Ward's 1898 'Improvements in Dress-Pocket Protectors against Pocket-Picking'.[28,29]

Charlotte Perkins Gilman's writings provide a particularly illustrative example of the political relevancy of pockets for women in this era. She was an American utopian feminist renowned for her novels, both fiction and non-fiction, and social reform activities. In 1914 she published a short story, 'If I Were a Man', which tells the tale of Mollie Mathewson, who awakes to find herself in her husband's body and proceeds to walk through the streets on the way to catch a train to work. She conveys the feeling of freedom that came from being dressed as a man.

> At first there was a funny sense of size and weight and extra thickness, the feet and hands seemed strangely large, and her long, straight free legs swung forward at a gait that made her feel as if on stilts. This presently passed, and in its place, growing all day, wherever she went, came a new and delightful feeling of being *the right size.*[30]

Mollie is surprised by how her new male legs move freely in trousers, unencumbered by layers of heavy petticoats and long skirts, how her waist felt free of a corset, and her feet comfortable in flat shoes. She does not need assistance to get about. Profoundly, she finds that she *fits* in the urban landscape.

> Everything fitted now. Her back snugly against the seat-back, her feet comfortably on the floor. Her feet? ... His feet! She studied them carefully. Never before, since her early school days, had she felt such freedom and comfort as to feet – they were firm and solid on the ground when she walked; quick, springy, safe – as when moved by an unrecognizable impulse, she had to run after, caught, and swung aboard the car.[31]

In her husband's clothed body, Mollie interacted with the urban landscape and different mobility technologies in ways she had not done since childhood. There was a connection between her body and the footpath, door handles and the train seat. The masculine garments offered her a great sense of physical and also ideological independence. Most strikingly, she meditates on the politics of pockets in the costume.

> These pockets came as a revelation. Of course she had known they were there, had counted them, made fun of them, mended them, even envied them; but she had never dreamed of how it *felt* to have pockets.[32]

Charlotte Perkins Gilman's character was able to inhabit and negotiate the urban mobile landscape differently in her husband's clothed body. Being able to control her body in new ways delights her. In this story, clothing, and the pocket more specifically, is a powerful metaphor for the frustrations women were feeling about barriers to public space, money and property, education and the right to vote. In this context, it is not hard to see why Henrietta felt the inclusion of multiple pockets was so critical to her costume for cycling.

Interviewing Henrietta's Three-Piece Convertible Cycling Suit

Henrietta's patent provides detailed instructions to produce a three-piece suit suitable for walking or cycling – including a long tailored coat, convertible skirt and a combined vest and knickerbocker. As per the other costumes, making the suit involved working with a pattern cutter to produce a block pattern inspired by the patent, producing toiles of each item in a similar weighted wool to work through the design, before constructing and wearing the final piece.

Upon first impression the patent seems complicated. The drawings demonstrate different elevations and assemblies which make the suit look even more intricate than the detailed textual description. Added to this is the sheer volume of pattern pieces needed to make three garments – there were 37 pattern pieces in total and the most of all the garments in the collection (it could have been slightly less but we added lining to the coat and skirt). Yet, upon making Henrietta's invention what emerges is its simplicity. The design strips away extraneous materials and details, and concentrates on what is essential – a garment that has no unnecessary frills or layers, is light, easy to put on, simple to convert, and enables a woman to move relatively freely. Henrietta's commitment to progress for women is materialised in this garment.

Historical and genealogical research revealed how Henrietta approached the idea of women's freedom of movement from every angle – health, education, marriage, reproduction, employment, media, politics, religion and beliefs, travel and the body. If in doubt she declared her aims in the *Woman's Herald*: 'Our readers know that the aim of the paper is to further the emancipation of women in every direction and in every land.'[33]. A similar drive emerges in her patent. 'These improvements consist in the form and combination of three specially constructed articles

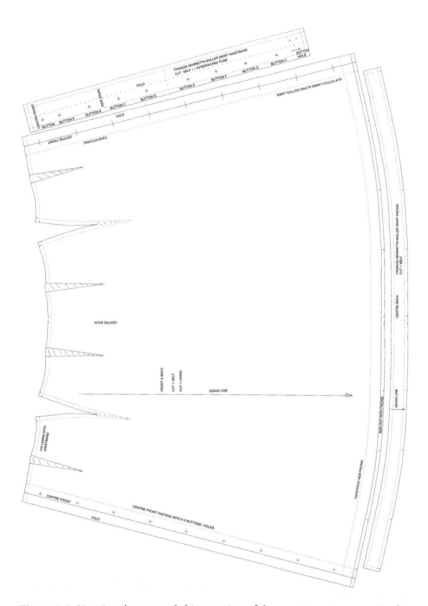

Figure 8.6 Henrietta's patented skirt consists of three pattern pieces; main skirt body, hem facing and waistband

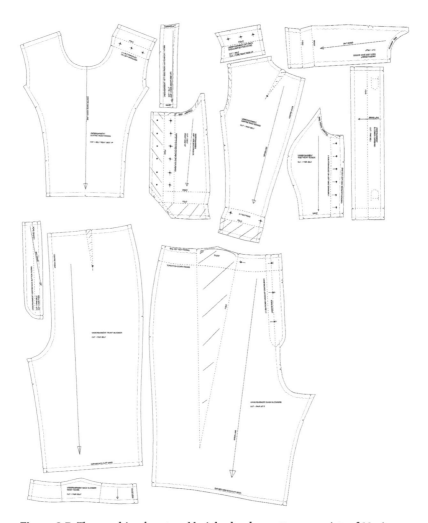

Figure 8.7 The combined vest and knickerbocker pattern consists of 11 pieces; centre front bodice, centre back bodice, side front bodice, left side back bodice, right back side bodice, cuff band, bloomer back, bloomer front, centre back button facing, side front button facing and waist facing

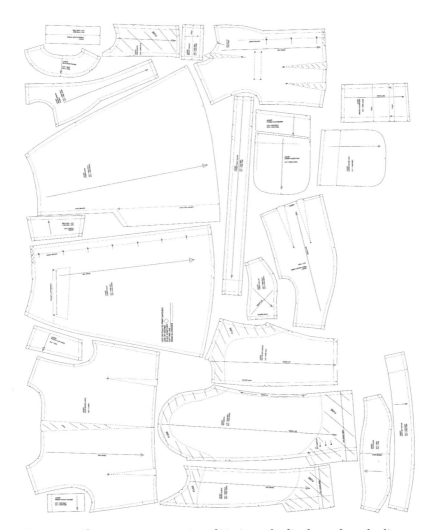

Figure 8.8 The coat pattern consists of 23 pieces; bodice lower front, bodice lower back, centre back bodice, side bodice, back bodice facing, front bodice facing, front bodice lining, back bodice lining, top collar, under collar, top sleeve, under sleeve, breast pocket, breast pocket bearer, breast pocket bag, coat pocket bearer, pocket flap, lower pocket bag, pocket flap lining, back bodice lining, front bodice lining, lower front facing, back neck facing, lower placket piping

of ladies' costume, so made as to afford special facility and convenience when cycling.'

Other costumes in this book suggest ideas for complementary garments or fabrics, but generally leave details and accessories up to the wearer's discretion. Henrietta does not. She addresses the entire suit as a system. Each piece is carefully considered and streamlined to minimise impact on the body. This is particularly evident with the undergarment. Here, two conventional items – blouse and bloomer – are replaced with a combined sleeveless vest and knickerbocker garment. The fact Henrietta has even focused on an undergarment is interesting. While this was a garment not normally seen by the public, the virtues of 'Rational Underclothing' was a subject of discussion in *The Rational Dress Gazette*. Members advocated lighter, looser and less cumbersome styles and fabrics and viewed undergarments as equally important as other items of clothing. 'It must not be forgotten, in speaking of rational dress, that rational underclothing should be one of its chief features.'[34] A combined garment was one way of dealing with heat and irritation caused by wearing layers. Buttons at the neck and arm and also at the waist and knee keep it in place. There are also buttons at the lower back which enable wearers to relieve themselves without removing the entire garment. There is no mention in the patent for the wearing of a corset. Henrietta's combined vest and knickerbocker also removed the issue of having to contend with a double waistband (this might be a reason why the coat is also without a belt).

The skirt, like others in this collection, is based on an ordinary A-line design. However, this version has a centre front seam, which is closed with nine buttons and buttonholes. Darts at the waistband shape it to the wearer's body. Like the rest of the suit, Henrietta sheds as much unnecessary detail as possible. She suggests a shorter length skirt to 'clear the ground say by about six inches'. She also recommends '[d]ispensing entirely with the lining' and avoiding any fancy edgings, by putting into place 'only a narrow stitched hem around the bottom'.

The convertible mechanism in this patented design is a clever but simple system of ribbon loops and buttons. Six loops are sewn inside the skirt hem in line with corresponding buttons on the waistband. To convert the garment, the hem is hooked up to the waistband via these buttons and loops. Depending on how many loops are fastened, the entire skirt can be lifted from the ground, or just the front. The coat is knee-length with ten buttons to close. There is also flexibility in how it can be worn fully buttoned or half-buttoned. Extra buttons are included so that the front panels can be fastened back to allow for even more freedom of movement.

Henrietta's looped skirt can be compared to Julia's gathered semi-skirt in terms of the inside-out nature and the radical change it enacts from ordinary wear. However, unlike Julia's, it is not on show. When looped up to the waist the coat entirely conceals the skirt. Henrietta explains: 'By continuing the looping process this garment can be entirely hidden under the coat', which meant 'a machine of the diamond frame type could be readily ridden.'

Hernietta's costume is the most formal of all the designs in this book. It would not have looked out of place in a parliamentary debate or public lecture. It is not a complex design, but one born of experience and interest in making things better. It preserved what Henrietta considered essential and removed any excess. Her aim was to give 'flexibility and freedom to the wearer'. For Henrietta, it seemed that clothing was as much a device of freedom and progress as the printing press, protest or bicycle itself.

The only significant addition Henrietta makes to the otherwise pared-down design is in terms of pockets. To make five (or more) pockets in Henrietta's suit provided time for us to reflect on pockets as gendered devices of mobility, as discussed above. While Henrietta makes clear the nature of the pockets, the drawings do not provide instruction for specific styles. Yet, each is different – in location, construction and purpose. The nuance is left up to the maker. Two in the coat were bigger, and the easiest to access in the context of the complete suit. These pockets were on show – a demonstration of power. We made them slashed pockets with welts and flaps and positioned them at the hip. The vest pockets would have been smaller, given they were for a (pocket) watch and the like. The skirt pockets were medium-sized and placed so as not to hinder the movement of the legs or comfort on the saddle. What became evident through sewing and wearing the costume is how the converted skirt itself creates more pocket-like spaces. When the front is looped up, the material creates two large enclosed spaces, and when the back is joined at the waist it makes even more. Henrietta refers to it as a 'Fishwife' skirt. This was a full double-layered style of skirt worn

Figure 8.9 Converting Henrietta's three-piece cycling suit

Table 8.1 Requirements

Suit materials	Extra garments
6m Dashing Tweeds – Urban Check	Boots, tights, scarf,
3m mustard linen blend	hat
6m mustard and purple rickrack braid for trimming	
2.5m printed research lining	
14 buttons for combined vest and knickerbocker	
15 skirt buttons (9 centre seam and 4 side opening)	
10 coat buttons	

by a workingwoman in Scotland's fishing industry, whereby the upper layer could be caught at the waist to create holding capacity. Henrietta's entire garment is in some ways a giant pocket.

Overall, what emerges in this chapter is how Henrietta was not interested in addressing a single barrier to women's emancipation, but rather in approaching all of them systematically. She was sensitive to the restrictions on women's lives, having fought on many fronts for her own and others' rights to gain an education, the vote, reproductive freedom and equal pay. Women's rights activists knew that they needed to dismantle the system, to change lots of things, not just one, in order to achieve true freedom and equality. It makes sense that she would also focus her attentions on an entire assembly of clothing for newly mobile women.

9

Patent No. 13,832: Mary and Sarah Pease and Their Convertible Cycling Skirt/Cape

Mary Elizabeth Pease and Sarah Anne Pease submitted a complete specification for 'An Improved Skirt, also available as a Cape for Lady Cyclists' on 5 March 1896 and their patent was accepted on 11 April 1896. They were residing at Sunnyside, Grove Road, Harrogate, Yorkshire and, like Henrietta, they identify themselves as gentlewomen.

The Pease sisters are two of the younger inventors of the period, aged 23 and 24 at the time of their patent, and the design reflects their exposure to a changing society. They recognised the popularity of rational attire for cycling women, yet they were also acutely aware of how wearers could be the target for scorn and abuse. This was often at times when the woman stopped cycling and had moved away from her bicycle. They explain: 'The rational dress now greatly adopted by lady cyclists has one or two objections inasmuch when the lady is dismounted her lower garments and figure are too much exposed.'

This garment is one of the more radical designs of the time, because the skirt completely comes away from the body. The sisters 'improved' the skirt by making it detachable and giving it a dual purpose. It is both a skirt *and* a cape. If made in a light fabric, it could be rolled up and attached to the handlebars or waistband using the gathering ribbon. The sisters identified a market for a garment that enabled women to conform to or challenge social conventions at times and places of their

Nº 13,832   A.D. 1895

Date of Application, 19th July, 1895
Complete Specification Left, 5th Mar., 1896—Accepted, 11th Apr., 1896:

PROVISIONAL SPECIFICATION.

Improved Skirt, available also as a Cape for Lady Cyclists.

We MARY ELIZABETH PEASE and SARAH ANNE PEASE both of Sunnyside, Grove Road, Harrogate, Yorkshire, Gentlewomen, do hereby declare the nature of this invention to be as follows :—

This invention relates to an improved skirt available also as a cape for lady
5 cyclists.

The rational dress now greatly adopted by lady cyclists has one or two objections. inasmuch that when the lady is dismounted her lower garments and figure are too much exposed.

Now the object of our invention is to obviate these disadvantages by the
10 formation of a cape like garment which can be secured round the neck as a cape or be secured round the waist as a skirt and occupy a position under the usual "fall" of the tunic but to a lower level. The garment can be readily put on by first suspending the front from a hook or other fastener.

Our cape and skirt garment is preferably made of a light or thin waterproof
15 fabric of the desired pattern to coincide with the colour or texture of the dress, it is of circular shape with a clear opening at say the front, to be secured by buttons hooks and eyes or other fasteners in its depth to overlap and with a band in, which a running cord or tape is arranged for drawing the band in to the required. diameter or size of waist when worn as a skirt and of the neck of the wearer when
20 worn as a cape the band being so made that when drawn in for the neck it gives the appearance of a Ruché collar.

The duplex garment thus made can when not in wear be rolled for attachment to a cycle or be folded and be carried in the cycle pouch or it can be suspended in a. rolled up condition from a lady's waistbelt.

25   Dated this 19th day of July 1895.

H. GARDNER,
Patent Agent, 166 Fleet Street, London,.
Agent for the said M. E. & S. A. Pease.

Figure 9.1 Excerpt from Mary and Sarah's convertible skirt/cape patent

choosing. They explain: 'Now the object of our invention is to obviate these disadvantages by the formation of a cape like garment which can be secured round the neck as a cape or be secured round the waist as a skirt and occupy a position under the usual "fall" of the tunic but to a lower level.' The skirt/cape's dual identity points to the tensions implicit in the desire to participate and support a progressive new order and at the same time not completely reject social propriety.

This chapter explores cycling's changing visual culture. Cycle wear radically transformed how women's bodies looked and the speed at which they moved in public space. Cycling women were outside, moving much faster than pedestrians and sometimes alone, without a chaperone in urban public spaces. New styles of cycle wear in the form of bifurcated garments – such as bloomers, knickerbockers or short divided skirts in place of ordinary floor-length skirts and loosened or no corsets – produced a very differently shaped female form. They stood out from other women of their class and sparked initial shock as well as larger broader debates about women's role in society. Unlike Henrietta, there is no evidence that the Pease sisters were women's rights activists, but their contributions still played a role in shaping this movement. As Fiona Kinsey notes:

> The safety bicycle made cycling more widely accessible to women. These women cyclists were typically white, and middle- and upper-class. Such women, while not necessarily suffragist them-selves, were at the vanguard of changes in gender roles which ena-bled them to take up new forms of physical activity.[1]

Many of cycling's early adopters were also at the vanguard of other new emerging technologies, such as the camera. Being associated with both in some form displayed technical competency, social status and cultural capital. Together they enabled women to embody and repre-sent different forms of public mobile femininity and played a role in changing the visual landscape of gendered mobility and citizenship in Victorian England.

The Inventors and Their Lives

Co-patentees were rare. Sole inventors are the norm, even if the reality of the design origin is far from independent production. Mary and Sarah were sisters and close in age, born 1873 and 1872. They were living at 'Sunnyside', their family's farming estate, which according to the 1881 Census, it was a substantial sized property at 209 acres and employed three men and two boys. Mary and Sarah's mother, Elizabeth Annie Pease, managed the estate. She was listed as head of the household and a widow. Her husband, a farmer, John Pease, had recently died, in 1879. The sisters had two other siblings – an older sister, Minnie, and a younger brother, John Cockcroft. All were listed as 'scholars'. The Census

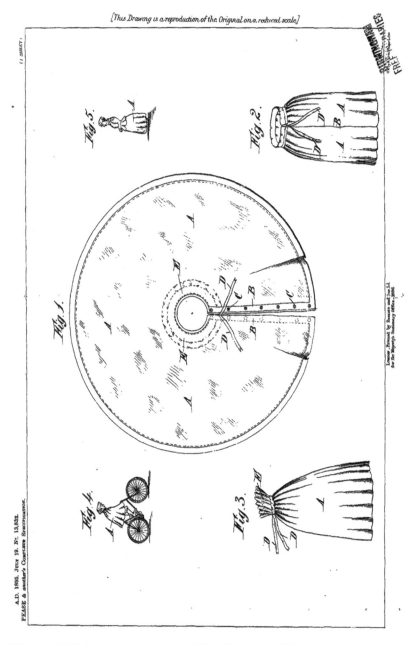

Figure 9.2 Illustration from Mary and Sarah's convertible skirt/cape patent

on 3 April records a further seven people in the house (which may have been in addition to the workers above): a governess, Eleanor Stroughaim (26); four servants, Mary Clarkson (22), Polly Ogle (16), Richard Warrior (17) and Henry Edmundson (16); a shepherd, Thomas Dixon (72); and a visiting merchant, Abraham Cockcroft (36). Given the latter shared the same name with John, the son, he was likely a close friend of the family. The size of the estate and household explains why the sisters self-identify as gentlewomen.

A decade later, Elizabeth Annie Pease was still the head of the household at 52, and 'living on her own means'. Only two daughters are at home now, her eldest, Minnie (23), and Mary (18). Sarah was 19 and had a job working as a governess with a Yorkshire family, headed by a schoolmaster. The family farm 'Sunnyside' was still a good-sized household, but had reduced in scale. On the night of the 1891 Census, Sunday 5 April, there were three other people in residence: a family visitor, Caroline Barclay (53) also living on her own means; and three servants; a stable boy, William Jupiter Borthwick (14), a domestic servant, Elizabeth Teny (21) and a shepherd, George Downes Soothill (45). Given there are no other employees, its possible that the family had disposed of the larger acreage and were living on the proceeds.

Were the sisters cycling together, even though one was living and working away from home? Sarah was still in Yorkshire, so it is possible that she was able to return home for occasional visits. Were they sharing ideas from periodicals, sewing and talking about things that Sarah had seen in her new job? We can't know for sure, but we can draw on experiences of other women engaged in similar pursuits. First-person accounts by women about women's cycling experiences are rare. There are more accounts about women by others in the public record. However, there are some illuminating exceptions that we can use to gain a sense of what the sisters might have been doing and thinking. The letters shared between Kitty and her companions, for instance, provide insights about the kinds of cycling costumes women at the time were wearing, the adventures they had and the kinds of responses they were getting from fellow road users. On 13 September 1897, Kitty wrote to Maude:

> It was market day on Friday at Chippenham and we created quite an excitement, though I think as many looked on with approval as

Figure 9.3 Our version of the Pease sister's patented convertible skirt/cape
in action

those who laughed and whooted. Anyway it was a good natured
crowd and nothing to hurt was yelled. The way we steered up the
street, through sheep, cattle and farmers was fine. The latter took
no notice of the bells and did not budge until the front wheel was in
their backs. Devises seems an uninteresting town and the waiting
maid who served us with tea at 'The Bear' held her head in the air,
sniffed and would not speak more than was necessary.[2]

Another first-person account even more closely located to the sisters' home in Yorkshire is found in Emily Sophia Coddington's diary. Emily was of a similar age when she took up cycling around the northern area of Leeds, Harrogate, Collingham, Ilkey and Dynley. She kept a richly detailed account of her new hobby from June 1893 to July 1896.[3] It is a remarkable account as she rides at least a few times a week, sometimes twice a day, and keeps meticulous record of her excursions in a neatly organised book, complete with columns headed 'Places', 'Date', 'Miles', 'Weather' and 'Comments'. She cycled through all seasons, in hilly conditions and, as the diary attests, for increasingly longer distances.

> Harewood, Collingham, Shadwell and E. Keswick to Chapeltown and Shaw Lane, home. June 24th 1894. 5.30–7.30pm. 22. Beautiful eve. Wind still high. Very hilly ride, went down the hills at a tremendous pace, like an engine. H. could not keep up with us. It was glorious.

Emily tallies the distances every year. In 1894 she cycled 1,075 miles and then a year later it was 1,458.5. Her diary tells the reader about the landscape, which was hilly, sometimes windy and often muddy, but for Emily and her bicycle it was filled with 'quaint small towns', exciting downhills and beautiful vistas which in all weathers made the effort worthwhile. Emily spends less time describing her costumes for every outing but there are still some clues. On 8 April 1894 she writes that she wore a new costume that had a 'rational underneath' and which was beneficial 'especially when the wind blew'. This most probably was some form of bloomer or knickerbocker in place of petticoats. Then, on 6 September, during a 'fine but windy and rainy' evening on a road that was 'very hilly', she almost had a crash while cycling. 'Very nearly had a spill, dress wound round the pedal', which led her to exclaim, 'why don't we wear rationals??' Although she recognised the benefits of wearing less conventional dress while cycling she continued to wear skirts. Emily may well have welcomed the Pease sisters' convertible skirt and cape design.

The sisters registered their patent in 1896 and even though only one of them was still living at home, they list 'Sunnyside' as their residence on the patent. At the turn of the century, both daughters had left home. Only Minnie was living with her mother, which was unusual as she was the eldest. The presence of a hospital nurse, as well as a servant, living in the house on the night of the 1901 Census suggests that either she or her mother were not well. Interestingly, one of their closest neighbours was a retired patent agent, Edward Dutton, who might well have played a role in inspiring or encouraging the sisters with their invention.

Like many women of this time, there is little record of the sisters' lives apart from Census and Birth, Death and Marriage records. Even with this, we cannot be absolutely certain that this is the exact Mary and Sarah Pease who lodged the patent for a convertible cycling skirt and cape in 1896. There is also no evidence that their patent was licensed and produced in their name. However, we can still hear their voices in the patent and see their ideas in a design that points to a range of issues prevalent to newly mobile women at this time.

Tactics for Site-Specific Concealment

Mary and Sarah's approach to the 'dress problem' is very different to their contemporaries. In an attempt to get the skirt out of the way of the wheels, they offer an option to remove it entirely. While the sisters were aware of the benefits of cycling in some kind of rational dress such as bloomers or knickerbockers without a skirt, they were also mindful of when it might be safer to have a backup plan. They were not alone. Even for confident cyclists, like Emily Sophia Coddington, who otherwise embraced cycling as part of their social identity and everyday life, cycling without a skirt was a risky move.

Popular columnists like Marguerite, who penned 'Lines for Ladies' in *Bicycling News*, advocated the carrying of skirts for when the wearer needed to get off the bicycle and return to more perambulate socialites. But even then, these preventative measures did not always guarantee a pleasant experience.

> We both carried our skirts on our machines so that we could slip them on for walking, and at the hotel where we were to stay for the night. We encountered plenty of adverse criticism and some rudeness in almost every town and village through which we passed, but for downright insults the manhood of St. Alban's was pre-eminent. The remarks which were hurled at us as we passed through that seemingly respectable town were such as would make any woman,

however strong-minded, make a resolve that nothing should in future tempt her to venture to ride in a costume which laid her open to such insults ... it is hard to break through so much prejudice, and the effort calls for much more courage than the average woman is possessed of.[4]

Although keen to ride occasionally without the encumbrance of the skirt, but possibly because of the adverse social response it catalysed, Marguerite was certain that the time when 'the last lady cyclist will "cast aside" her skirt is as yet far distant'.[5] The persistent presence of the ordinary skirt became such a common trope that it was gently mocked in women's magazines and cycling periodicals:

> Maud: 'Do you wear the rational dress?'
> Gertie: 'Yes, but I cover them with a skirt'.[6]

Carrying a skirt for walking away from the bicycle fits with the site-specific strategy outlined in Chapter 3. Women would wear one costume to do city cycling and another for 'proper' longer or faster rides. The latter was often without a skirt away from social scrutiny. There were many designs at this time that aimed to provide something in-between for cyclists of different active and political persuasions. Under the banner 'A Costume for the Country', London's *The Morning Leader* suggested the following convertible costume was appropriate for cycling as well as other activities like golfing or country wear:

> The bodice is double-breasted, and it fits closely down the back and side, though ponching slightly in front. It can also, you see, be buttoned at the throat. Neat breeches should be made to wear under the skirt, with a band of box cloth below the knee, and if you are sufficiently of the 'new woman' you can make it into a rational costume by discarding the skirt and donning a long basque-like garment of the same material, which buttons round the waist, is unlined, and reaches only to the knee.[7]

Mary and Sarah's patent is similar but goes one step further. It gives the skirt new purpose, as another garment entirely – a cape. It was designed with a quick transformation in mind. As they explain, the aim of the design was such that 'on dismounting if the article be in wear as a cape its removal and securing round the waist would in a few moments convert it into a skirt without making the wearer unlike others in the vicinity'. It would have made an ideal garment for Kitty's friend Minnie:

> Minnie came from Harborough part of the way by train the rest cycling, she got in late for she as quite done up by the heat. K.W and self wore no skirts on Sunday, some friends of hers came to tea and she wanted them to get used to the costume. It was jolly wandering around the woods without a skirt and Minnie wished she had her costume. But she wore a skirt because of going by train and did not bring a coat.[8]

Cycling women generated a great deal of attention. While some attempted to avoid it as much as possible, for others it was unavoidable and for a few this attention worked in their favour. Tessie Reynolds, for example, generated a huge media response to her long-distance cycling achievement as well as for her rationally inspired handmade costume. This may well have inspired many to wrestle with their own anxieties. S. S. Buckman, writing under his pseudonym in *The Lady's Own Magazine*, was adamant about the importance of the press for the Dress Reform Movement:

> The great thing in connection with Rational Dress is to obtain publicity. Only by so doing and by familiarizing the public with the idea of the costume, as well as letting them see the costume itself, with the great desideratum be soon attained – that the public accept the wearing of the dress as a matter of course.[9]

Lady Harberton, rarely accused of being publicity shy, sought to further the cause of dress reform through media attention, public debate and also a famous court case when she was denied entry to the Hautboy Inn coffee room in Surrey for wearing rational dress. She was firm in her conviction that dress reformers needed to used their differently clothed bodies to change public opinion. She encouraged women to wear their costumes at all possible opportunities to claim this modern identity as a new visual norm. 'It is difficult to over-estimate', she declared, 'how much may be done in home life and private social ways'. Having said that, she was still sensitive to the fact that some members may well be fighting resistance at home as well as in the streets. But overcoming even domestic opposition was a step forward. 'By having the courage of her convictions, the Leaguer will find that she can generally arrange to wear the dress some part of the day in her lodgings or boarding house, flat or home and so the prejudice of parents and brothers, friends and fellow-boarders may be overcome, or at least the way made easier for others.'[10] She was, however, less enamoured by the skirt-carrying method:

> We have never thought it necessary to carry a skirt on our machine. It makes a cumbersome parcel, and what good does it do? We earnestly hope those wearers of rational dress who have hitherto carried, and occasionally put on skirts, will discontinue the practice.[11]

While there was much encouragement to wear new forms of cycle wear and influence the view of the general public, it was not easy. Some cyclists like Kitty and her friends relentlessly just kept doing it, putting their differently clad bodies on the front line, and coped with the responses as best they could. Others sought to claim more private spaces first. In 1898, Miss E. Whittaker declared herself a member of the Indoors Reformed Dress League. Interviewed in the *Daily Mail*, she made clear that she did not cycle and chose instead to showcase her bloomers at home and claim this space first before venturing out onto the street. Miss E. Whittaker argued that 'the cause' was being harmed by outdoor dress reformers. She strongly felt the home was the first challenge for women dress reformers before seeking to claim outside space.

> The object of the league is to make reform easier by avoiding the publicity of an out-door reform. The founder thinks that hitherto a grave error has been committed in trying to force a reform upon a public which is not educated up to it ... When this Indoors League gains all the recruits it hopes to secure ... Then shall the lady cyclist who merely wears her 'rationals' when wheeling feel what a spurious, half-hearted reformer she really is.[12]

Despite Miss Whittaker's misgivings, there was space (and interventions) for all kinds of resistance. The sisters' skirt and cape offered women the opportunity to claim different mobile identities at times and places of their choosing. The patentees seem to understand the importance of being part of the new visual culture of the time, to support women participating in this exciting vision of progress and modernity. However, they were also aware of the trouble this might catalyse and a woman's desire to have at least some control over it.

Visual Culture of Women's Cycling

Victorians were quick to embrace not only the bicycle but also many other new and exciting technologies, such as photography. As Charles P. Sisley, editor of *The Lady Cyclist*, explains:

> There is no doubt that touring combined with photography is the most enjoyable way of spending one's holiday, and although I have not tried the experiment of snap-shotting myself I know very many ladies who have, and they are delighted with the idea. Last Saturday, when taking a ride to Epping, I saw no less than four cyclists who had a small bracket attached to their machine for the purpose of carrying cameras.[13]

In addition to touring photography, the fusion of these two pursuits resulted in the advent of cycling portraits. These images reinforced the subjects' cultural cachet on multiple fronts – it was a social achievement to be able to afford access to new cutting-edge technologies, be associated with these symbols of modernity and demonstrate new skills required to participate in these cultures. Adverts for new costumes, bicycles, sewing machines and cameras were regular features in popular cycling periodicals. Portraits also accompanied columns in *The Lady Cyclist* such as 'Lady Cyclists at Home' and 'Why a Lady Cyclist Should Always Dress Well'.

Bicycle portraits took a number of forms. Cyclists of all ages from young girls through to matrons were photographed with their bicycles, often alone and un-chaperoned and in front of a natural landscape scene. Most were studio based, though some were taken in outdoor sites. Regardless of whether the backdrop was real or painted, the cyclist was positioned in an outside context, which presumably she was about to cycle away into. The range of bicycle styles and costumes is vast – from conventionally skirted riders who sit upright on step-through frames to scorchers bent forward over diamond frames in close-fitting knicker-bocker suits. Irrespective of whether the chosen garment was practical for cycling, they symbolised a woman's confident engagement with multiple new forms and sites of modernity – the camera, bicycle, modern dress, independence and public space.

Bicycle portraits were encouraged by popular periodicals. Many magazines ran competitions for the best new cycling costume. A 1986 competition run by *The Lady Cyclist* stated: 'All intending competitors have to do is to send a photograph of themselves, either mounted on their bicycles or standing beside them, in fact, in any position they prefer, provided they are in cycling costume. The costume may be either of the skirt or knicker pattern, and the prettiest picture will gain the prize'.[14] What was deemed 'prettiest' in these kinds of competitions was largely dependent on the political orientations of the publication. In the case of *The Lady Cyclist* competition, the fact they welcomed both skirted and non-skirted riders opened up the contest to a broad readership. The 'novel prize' was somewhat ironically cycling accident insurance but there were lots of consolation prizes (for 12 runners-up), which is indicative of their larger aim – to generate a visual gallery that would help to sell more copies of the magazine and in doing so, support and normalise the wearing of a range of cycling wear.

One way to change public opinion about women's cycling was to normalise it through ubiquity and familiarity. The *Rational Dress Society* knew this well and it encouraged members to not only wear their costumes at every possible opportunity but also to actively participate in the visual culture of cycling by having their portraits taken, thereby forging cycling imagery in the public domain that featured not only men's bodies, but women's as well. In fact, members were encouraged to take up an offer by a local cycling photographer:

> Mr Clare Fry, of the well-known firm of photographers, C.E. Fry & Son, 7. Gloucester Terrace, S.W., is a member of the League and has generously offered to photograph free of charge any member of the League and present her with one set of finished proofs.[15]

Why was this important? Some have argued that the bicycle portraits played a critical political role by presenting alternatives to dominant negative public imagery. Bicycle portraits represented carefully curated and positive images of women cyclists at a time when they were consistently 'caricatured by the media as masculinised and unattractive'.[16] As discussed earlier, some parts of society saw them as having relinquished their femininity, by embracing masculine behaviours and clothing. They were often portrayed as poor cyclists not in control of their skirts, hair or sweat glands, let alone their velocipedes. Bicycle portraits provided a critical means for women to contest these representations.

These staged images would have been judiciously planned, shopped and sewn for, and set up. Postures are confident and in some cases daring. Women look directly into the lens, and at viewers. This can be seen as a precursor to what Lisa Tickner has written about in her study of *Imagery of the Suffrage Campaign 1907-14*. Her research explores the role of women artists in contributing suffrage imagery to the public sphere about women's lives, rights and emancipatory intent. These women recognised the power of popular visual culture 'in the maintenance and reproduction of anti-feminism.' Using the same tools and platforms they sought to counter negative representations with a positive and broader range of imagery. 'Feminists were regularly caricatured as over- or under-sexed, ugly, hysterical, masculine or incompetent.'[17] Much like women cycle wear patentees, artists used skills, bodies, materials and spaces available to them to intervene in this visual landscape and offer alternative ways of seeing and thinking about women in public space.

Looking closely at late-nineteenth-century bicycle portraits also reveals evidence of various degrees of enhancement – coat buttons and pleats are keenly defined, hat feathers are sharp, wheelspokes are outlined and steel rims have sparkle. Touching up was a common photographic practice in the Victorian era (as it is today). In this context, this practice further enhanced the attractiveness of the rider and overall appeal of the image. It is no wonder that bicycle portraits, as Fiona Kinsey writes, 'became a particularly potent and desirable accessory.'[18] In some ways, they can be seen as a precursor to today's cycling selfie.

Many of the portraits we can still see today were published in popular magazines. There would have been many more commissioned for personal use. The resulting images, much like the labour invested in making them, would have been distributed in a carefully considered manner – sent to sisters and beaus or perhaps never sent anywhere, and kept instead as private symbols of freedom, courage and exciting possibility. It is not difficult to imagine that not everyone wanted to go outside or indeed, in less supportive families, even stay inside in new cycling garments. Having a portrait taken was another way of participating in this new cultural practice. Cycling portraits would have enabled many women to envision themselves beyond domestic thresholds that defined everyday life. In all forms, they contribute to the rich and vivid visual culture of Victorian women's cycling.

Figure 9.4 Cycling portrait of Mrs Houston French, *The Cycling World Illustrated*, 1896

Figure 9.5 Cycling portrait of Mrs Barrington, *The Wheeler*, 1894

Figure 9.6 Cycling portrait of Mrs Selwyn F. Edge, *The Lady Cyclist*, 1896

Figure 9.7 Portrait of cycle racer, *The Sketch*, 1896

Interviewing the Pease Sisters' Convertible Skirt/Cape

This is by far the simplest garment in the collection. There is no hidden pulley system, no hooks or loops, and it is not part of a multiple piece system. It is just a single full circle skirt with a wide band though which a ribbon is threaded so as to gather at the waist or at the neck. Technical specificity in this patent is low, suggesting the sisters may not have been well skilled in dressmaking. But this does not negate its inventiveness; rather it just means that their focus was on bringing to life their idea through fashionability, flexibility and concealment.

The first unusual feature of this garment lies in how the sisters have not started with a conventional A-line skirt. Their version is shorter and fuller than a standard floor-length garment, and based on a circular pattern. This would not look as 'ordinary' when operating in skirt mode as per the designs of Alice, Julia or Henrietta. However, shorter skirts were becoming popular with cyclists, as shown in periodicals such as *The Ladies' Tailor*, so it would have looked like a new cycling garment when worn over bloomers or knickerbockers. It is also unique in that it is not tailored to a single body shape. All the other skirts in this collection were made to fit the wearer. In this case, the ribbon or cord at the waist gathers to fit. As such it accommodates a broader range of bodies.

The complexity of the patent lies in the cut. It is not just a full skirt. It is also a cape. The wide waistband doubles as a high collar, when lifted up to the shoulders. The sisters explain, 'Figure 3 [in Fig. 9.9] represents the skirt-au-cape, as a cape, the upper portion being drawn in at the neck to pleat the band portion, and make it look like a Ruché collar.' Upon loosening the gathered material, the collar then converts back into the skirt waistband. 'Figure 5 illustrates the article when worn as a skirt, the band portion and runner tape being hidden by the fall of the tunic and gives the appearance of the skirt of an ordinary walking costume.' The length of the skirt meant that it covered most of the body when worn as a cape. This invention is two completely different garments – cape and skirt. Neither item is compromised. They both work.

The sisters' emphasis on the ruche collar reflects a popular style of the period. There were many references available to help us make sense of this garment. High collars were routinely displayed and discussed in women's fashion magazines such as *The Queen*, mostly for spring capes

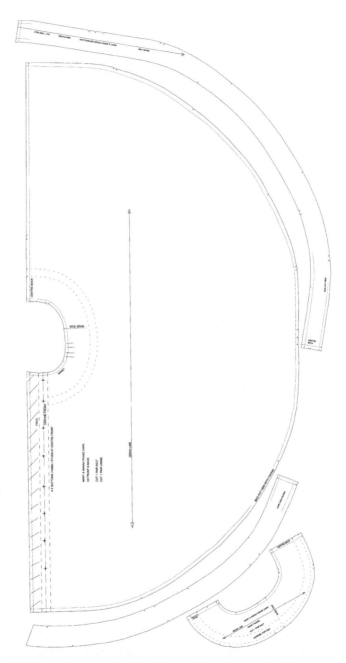

Figure 9.8 The pattern inspired by Mary and Sarah's patented skirt/cape consists of four pattern pieces: main skirt body (to be cut twice), two-part hem and collar facing (the hem can be left out to reduce weight)

Figure 9.9 Details of illustrations from Mary and Sarah's patent

and tailored coats. The inventors also paid particular attention to materials, more so than other patentees of the period. They recommended using a 'light waterproof or rainproof material', which would have made it a particularly practical garment in inclement weather, both as cycling skirt and cape. The lightness ensured it easy to fold up into a compact object to attach to the handlebars. So, in addition to providing site-specific options for wearers, it was also weather-specific.

Cyclists were early adopters of new technologies and materials and this is a clear example of the use of cutting-edge fabrics. Mary and Sarah also suggest using a 'material of reverse colours, say a check and a plain colour to suit or approach the usual colour of garments generally worn'. The purpose of this was to find an appropriate fabric match such that the garment would seamlessly become part of an overall costume regardless of how it was worn. The absence of extra buttons and fasteners, apart from those at the waist or neck, makes sense when read against this objective. Due to the fullness of the circular design, the skirt naturally falls shut without the need for extra closures along the opening edges. It also meant that a quick change between cycling and walking could be accomplished with minimum fuss.

Similar skirt and cape patents may have inspired the Pease sisters' design. American Alice Worthington Winthrop's patented her 'Bicycle Skirt' in 1895, the year before.[19] It is similar but a little more complicated than the Pease design, as it is *three* different garments. The skirt can be

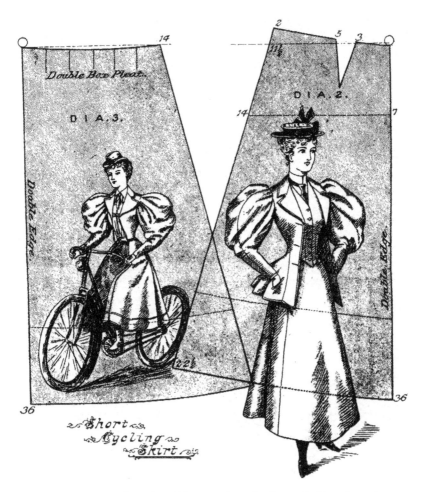

Figure 9.10 Pattern plates for a 'Short Cycling Skirt' provided by *The Ladies Tailor*, 1896

converted into a bifurcated garment in the form of a divided skirt via a series of buttons. The back and front panels can be removed and folded up to become a satchel attached to the handlebars or turned into a cape or a hood. Alice explains: 'In case the weather should become inclement or the rider become chilled, she dismounts her bicycle, detaches the rear portion of her skirt, and forms a combined hood and cape by drawing the ends of the cord through the rings having inserted her head in

Figure 9.11 Converting the Pease sisters' cycling skirt/cape

Table 9.1 Requirements

Skirt/cape materials	Extra garments
5m Dashing Tweeds – Navy Raver Wave	Waistcoat
10m navy bias tape	Blouse
5m printed research lining	Bloomers
2 buttons	Boots, scarf, tights, hat

the upper portion of the skirt.'[20] 'The most creative inventions', write Sally Helvenston Gray and Micheala Peteu, 'sought to provide additional utility to cycling attire.'[21] While there is no evidence that the Pease sisters' patent was commercialised in their name, there are many similar skirt/ cape garments promoted and sold by tailors around this time. So, it is possible that their patent was licensed, produced and sold under a different brand name. Despite our inability to trace their patent any further than the patent archive, what nevertheless emerges in their inventive practice is how aware women were of their visual impact on society and how they sought to work with and also press against these limitations through their clothing.

10

Patent No. 9605: Mary Ward and Her Convertible 'Hyde Park Safety Skirt'

Mary Ann Ward submitted a complete specification for 'Improvements to Ladies' Skirts for Cycling' on 8 February 1897 and it was accepted just over a month later on 27 March. The patent tells us her address but there is nothing about her married status or vocation. She was living at the time at 92 Thomas Street, Bristol, in the county of Gloucestershire.

The aim of Mary's patent, like the others, was to produce a garment that converted an ordinary skirt into a cycle-specific costume. The result was to be both functional and fashionable. This is the shortest patent of the collection; the complete specification is just over a page in length and it is lacking a technical diagram. Partly, this brevity is due to its simplicity. Mary's design is a full length A-line skirt fastened along the sides with buttons and buttonholes. The convertible system comprises two decorative straps sewn into the waistband that are tucked inside the skirt, hidden from view. When required, these straps are used to tether the skirt at intervals via the side buttons. This action gathers the sides of the skirt in a ruched manner, raising the material up and out of the way of the bicycle wheels.

This is not a radical transformative garment, like Julia's or Alice's designs. Archival research suggests this was not Mary's intent. Rather, she appeared to be interested in producing a convertible costume that was more subtle and site-specific for an upmarket urban client. This wearer wanted to look fashionable, and be safe, while undertaking social visits and public forms of city cycling. Mary was not overly enamoured with available rational styles or with wearing ordinary dress on bicycles and had the cultural cachet to push at the boundaries of conventional fashion that defined her class and gender. In her patent Mary explains: 'The utility of this skirt over the ordinary is, that whilst a stylish dress for walking may be worn, there is no danger, by reason of its length, in using the same when cycling.' However, this does not mean that the design was not located within a political framework. Mary garnered enthusiastic support for her patent from both the larger cycling and dress reform communities. She cleverly managed to design a garment that appealed to a broad group of women located between 'Skirted' and 'Rational Dress' supporters.

The theme of this final chapter relates to influential sites and communities where new modes of feminine mobility were being performed and produced. Even towards the latter part of the decade, there was still no single broadly accepted style of cycle wear. Women were in the process of working

N° 9605   A.D. 1896

Date of Application, 6th May, 1896
Complete Specification Left, 8th ·Feb., 1897—Accepted, 27th· Mar., 1897

PROVISIONAL SPECIFICATION.

Improvements in Ladies' Skirts for Cycling.

I, MARY ANN WARD, of 92 Thomas Street, Bristol, in the County of Gloucestershire, do hereby declare the nature of this invention to be as follows :—

-The skirt is conveniently fastened to the waist of the rider ; and from the part covering the hips to the bottom of the skirt is a division on each side, the edges of
5 each overlapping the other. On the edge of the under-lap is sewn a row of buttons ; and on the edge of the upper-lap is a row of holes. Over each of the divisions running down from the waist and secured at its upper end to the skirt is a strap which may be called a gathering-up strap in which also is formed a row of holes at suitable distances from each other ; so that when the two edges of the
10 divisions are buttoned together, the gathering-up strap may likewise be buttoned over the same ; thus affording a means of shortening the drop of the skirt to suit the wearer whilst riding. The skirt may be again let out to its full length for walking by merely unbuttoning the gathering-up straps. More than the two divisions can be made in the skirt if preferred ; and studs or links may be used
15 instead of buttons. Also the straps for gathering up the skirt may be secured apart from the skirt by means of a band or string if preferred.
The utility of this skirt over the ordinary is, that whilst a stylish dress for walking may be worn, there is no danger, by reason of its length, in using the same when cycling.

20   Dated this Fourth day of May 1896.

MARY ANN WARD.

Figure 10.1 Excerpt from Mary's cycling skirt patent

out what they should and could wear while engaging in this new means of moving in public. Opinions on novel forms of cycle wear were circulated, discussed and fiercely debated within the media, in cycling, fashion and dress reform communities and via formal and informal channels. Ideas were also being presented in three-dimensions, beyond text and talk, on the bodies of women who wore them. Mary's patented skirt was known as 'The Hyde Park Safety Skirt', which firmly located its use and users in a highly public arena. Places like Hyde Park, especially during popular periods such as the annual Season, were important sites where ideas around gender and class were performed, negotiated and re-made on a daily basis.

The Inventor and Her Life

Trying to gain a glimpse of Mary's life a century after she lived initially presented even more difficulties for the research team than Julia Gill. Her name is too common to even begin to come close to a firm match in genealogical records. She provides no hints of her vocation or married status. There is no evidence that her design was commercialised by a large fashion house like Alice's and her patent is brief and lacking illustration. Despite this there are still many clues for us to follow. Mary tells us where she lives and her design is easy to understand. There is also information to be found about the inventor and her invention in formal and informal channels of communication. The patent, news articles and personal correspondence reveal a great deal about the role Mary's design played in the development of new forms of cycle wear and, more broadly, women's freedom of movement in late-nineteenth-century Britain.

The first insight into Mary's patent comes directly from the Dress Reform Movement. Her cycling skirt attracted the attention of Lady Harberton. The design's merits were discussed in personal correspondence on 12 February 1898, between Lady Harberton and S. S. Buckman: 'I wonder if Mrs Buckman knows of the thing called "The Hyde Park Safety Skirt"', she writes. 'For it is an invention whereby the Rational Dress can be made into an ordinary looking skirt at once.' Lady Harberton recognises the convertible benefits and points out who might find it useful: 'It was made by a Mrs Ward and I have seen it and though I don't want

Figure 10.2 Our version of Mary's convertible cycling skirt in action

it myself, it might be convenient for anyone paying calls who wants to leave their cycle and walk about.' She then suggests the kinds of cycling that might be appropriate with this garment: 'It would not prevent a person riding a diamond frame.' She ends the missive with further clues for the researcher: 'It was described in *The Lady's Own Magazine* for December.' And just in case we were not already completely certain that this was our inventor, she confirms: 'The address of the inventor is Mrs Ward, 92 Thomas Street, Bristol.'[1]

Mary's skirt had been patented for just under a year when this letter was penned. The costume had been generating media attention during this time, which is why Lady Harberton wonders if S. S. Buckman's wife had an opinion on the garment. The article to which she refers is a two-page, three-photograph account of the garment in action. *The Lady's Own Magazine* was clearly as interested in this unusual garment too. In the article titled 'The "Hyde Park" Patent Safety Skirt', Mary is introduced to readers as a woman who 'for the past dozen years has been a keen follower of the pastime and a hearty supporter of the cause.'[2] Apparently Mary was a pioneering member of both cycling and rational dress communities in the West and South of England and committed to 'further the interests of wheelwomen and the Cause in particular'. She must have been a cyclist herself, as the article states that 'it is only those ladies who are actual riders who can be expected to know actually what is most comfortable and suitable'.

The article explains how the 'construction is simplicity itself; dispensing entirely with cords, tapes and elastics for alteration of shape'. This signals awareness of the range of convertible garments becoming available to cyclists at this time, including perhaps the 'Bygrave convertible skirt' and also attempts to highlight what is unique about Mary's design. 'It can be worn *a la* Rationals, and at the end of a ride immediately dropped into a full-length walking skirt; or it can be worn as a perfect safety skirt, being capable of adjustment to any height from the pedals.' We are told how the skirt 'can be made at a most reasonable price' and that it has been proven to work, as 'its practical use has been tested for close upon two years'. Mary must have tried out various iterations of her garment for at least a year before she patented it, and during the year subsequently. Interestingly, the article reports that Mary patented it to fill the existing gap for 'a really useful and workable dress'. The article goes on to describe even more advantages of the skirt beyond that of the description in the patent:

If on a journey, the wind is strongly against the rider, the divisions can be undone, and by buttoning crossways a short peak is formed in front and behind one upon which the rider sits, the other hanging down, both serving as shields to hide the fullness of habit natural in a lady, the disclosure of which, by the ordinary Rationals, is the very proclamation of their ugliness. If on the other hand, the wind is found to be strongly on the rider's back, the Dress may be worn skirt fashion, so as to receive its full, sail-like benefit – an advantage which even a gentleman cannot obtain.

Three photographs accompany the article to illustrate the convertible options[3]. We are not told who the woman demonstrating the garment is, but it could be Mary. Or, much like how Alice involved her sister-in-law Rosina Lane in the promotion of her patented garment, these images might be of one of Mary's sisters, as the article states that 'Mrs Ward, in conjunction with her sisters, has done much of the pioneering of the lady's safety in the West and South of England, and is ever ready to further the interests of wheelwomen and the Cause in particular.'

A year later the design was still regarded in high esteem. In April 1899, it was promoted on the front page of the *Rational Dress Gazette* where it was considered 'the most cunningly contrived thing we have seen.'[4] The skirt was recommended for 'ladies who wish to avoid the remarks made to rationalists.' Even at the turn of the century, women were *still* being verbally ill-treated for wearing non-conventional cycling wear in public. The *Gazette* described yet another useful feature of the garment – it could 'be adjusted on the machine in a moment as knicker-bockers, as a short skirt and as a long skirt.' Was this alluding to the skirt's flexible length, or something more? It is not entirely clear. Nevertheless, members of the Rational Dress Society were clearly supportive of the design, and as per their public commitment to connect inventors with members, they hinted that 'a pattern and illustrations of this skirt' might be made available to members in the future.

Promenading in the Parks

Why was the design called 'The Hyde Park Safety Skirt'? Exploring the name provides clues for developing a picture of Mary's influences and interests. Parks were primary public spaces in the late nineteenth century where the Victorian elite promenaded their new Safety bicycles and costumes. They were places to see and be seen in. In London, the main sites were Hyde Park in the city centre, particularly the Inner Circle and Rotten Row, which led from Hyde Park Corner to Serpentine Road, and Battersea Park in the west. Prior to the cycling boom, the parks were congested with lines of carriages carrying the social elite. An article in *The Queen* notes how this changed in the 1890s when the upper classes took up cycling:

> From the Achilles Statue to the Powder Magazine during the early hours the tinkling of the bicycle bell is the most dominant of all sounds, and the eclecticism of the riders is displayed in their way of riding, in their machines and their decorations, their toilettes, and in their general demeanour.[5]

As discussed earlier, many Victorian women employed site-specific cycle wear strategies, whereby different garments were worn for ordinary social riding and proper cycling. Park riding firmly fitted within the former and demanded particular attention to detail. Miss F. J. Erskine's *Lady Cycling*, written in 1897, provides hints and tips for 'what to wear and how to ride' for both town and country. 'For park riding', she counsels, 'we must have an artistically cut skirt, artfully arranged to hand in even portions each side of the saddle'. Miss Erskine is equally clear about what not to wear. She concedes that a fashionable 'blouse of silk or cotton, belaced, and with huge puff sleeves' is only suitable for good weather and 'as to their being any good beyond Battersea Park and the Inner Circle, the idea is absurd'. While more sensible clothes were appropriate for country riding, she was adamant that 'those riding in town must study the fashion of the hour'. Cycling in the park was not only for the pleasure of the cyclist. '[C]yclists can turn out so that it is a pleasure to see them'. These fashionable garments were not hard-wearing, but this was not their purpose, nor were Miss Erskine's readership the kinds of women who worried about such things, as she was well aware: 'These costumes will not stand wear, but those who ride in town can afford a change of dress for different surroundings, therefore consideration need be no drawback'.[6]

Cycling in Hyde Park was a relatively new practice in 1896. In February, *The Queen* reported that after much lobbying, cyclists were officially allowed to cycle there from eight o'clock in the morning until midday. 'The patience, the enterprise, and the frequently renewed efforts on the part of the society cyclists to annex some portion of Hyde Park in which to exercise the iron horse have at last been rewarded.'[7] This change in policy proved popular and thousands flocked to the park to enjoy the fresh wintery cycling.

A month later, despite the frosty weather, *The Queen* continued to enthuse about the popularity of the park, providing detailed description about the many types of cycling and cyclists who made use of the space. Apparently from eight o'clock through to ten o'clock, professional men used the park before heading to the city for work. Then from ten o'clock to midday, the park was filled with the echelons of society and spectators eager to glimpse the latest fashions. This was the late nineteenth-century's equivalent of celebrity culture.

> Hyde Park from ten to twelve every morning is quite sight. Whenever the weather is fine and the roads dry, from a couple to three thousand bicyclists ride up and down between the Archilles monument and the Magazine. Not only is there this vast concourse of bicyclists, but crowds of spectators line the side walks, promenading up and down, or standing leaning over the railings, to watch the passers-by![8]

Did it really get this busy? *The Hub* ran a regular feature called 'Pars from the Parks' which reported on happenings in London's parks. A park policeman was asked about the best time not to cycle but to 'properly see' what was called 'Cycle Row'. He suggested 'before breakfast, especially from 8 to 10 and about 11 o'clock'. He also declared that the biggest day they had during the 1896 Season was when 'about 370 cyclists passed along the Row'.[9] There is a significant difference between this and the 3,000 bicyclists mentioned above, but what we can glean is that Rotten Row was immensely popular for those on bicycle and on foot.

In fact, so popular was park cycling that it threatened to usurp conventional modal dynamics. 'A great grievance of Hyde Park

THE FATE OF ROTTEN ROW.

Figure 10.3 A satirical cartoon highlights some riders' shock at the changes in Hyde Park, *Punch*, 1895

Figure 10.4 People watching and cyclists chatting in Hyde Park, *The Sketch*, 1896

cyclists', declared *The Queen*, 'is that carriages come in large numbers to the already crowded road'.[10] A petition was submitted to the First Commissioner of Works in 1896 to allow cyclists sole access to parts of the park for two hours a day. This was in recognition that 'cyclists are now in excess of all other forms of traffic which frequent the park'.[11]

Activities in public parks were keenly observed, discussed and reported upon. *The Queen* noted that for the most part Hyde Park in early 1896 was filled with women cyclists wearing 'sensible short skirts and smart, tight-fitting little coats of tweed, serge, or clothing, with a neat Toque or hat to match', which appeared to pass the high standards of the writer. Less positive commentary was reserved for those challenging these conventions: 'Once or twice during the last few weeks has the serenity of the bicycling row been disturbed by an apparition in black satin knickers, surmounted by the incongruous widow's bonnet and crape-covered coat; the "skirtless" ones are not of these park riders, and it is to be hoped that they may long remain aliens to it'.[12] In contrast, writers in *The Lady Cyclist* saw little in the parks that warranted such harsh commentary: 'Cycling costumes are still commanding a great amount of attention ... Some of the riders in the London parks are noted for a little eccentricity in matters of dress, but the majority are both suitably and becomingly attired for the summer'.[13]

While some cycled through on their daily commute, parks were predominantly used for socialising. Illustrations show cyclists talking and waving to the crowds. Some are doing tricks, such as riding with hands behind their backs. Conversations were being had on bicycles as well as in the watching crowd. All look well-dressed with hats and gloves and other accoutrements. Cyclists' ages apparently varied from eight to 80, but the writer in *The Queen* notes how women made up the 'majority of the rapidly-riding midday crowd'.[14] Class was evidenced through clothing, choice of velocipede, posture and associated skills and technique:

> The company is entirely cosmopolitan, members of the best society, of all professions, and of the services being all en évidence. The ladies are naturally the more interesting of the wheeling votaries, and in all justice to them it must be admitted that they are by far the more pleasing riders, more especially in a place like the park, where

speed is by no means a desideratum, but where a graceful, becoming, and altogether correct seat is not only more enviable, but also much safer.[15]

Venturing into the parks to see the latest fashions and be seen by adoring (and critical) publics were not new practices. For the social elite, especially during the Season, public parks and squares were primary places to do social business. Historian Peter Atkins writes about how London, and particularly the West End, at this time was a 'container of frighteningly concentrated power'.[16] In May 1895, the *Tailor and Cutter* advised its readers that the 'best time to see art in gents dress in the park was Saturday afternoon between 3 and 5 o'clock' when 'both ladies and gentlemen make the park at this time a special meeting place'.[17] A year later, these same practices continued. What was new was the popular mode of mobility, which had shifted with cyclists replacing pedestrians, horse riders and carriages.

Park cycling may have been relatively new but it quickly secured a place in Victorian social life, such that many a social drama was played out against these backdrops. Battersea Park set the scene for the 1897 humorous story *Women and Wheels*, in which an aggrieved husband laments the loss of his wife to cycling.[18] The man's supportive friend has difficulty understanding the appeal of the park for women.

> The New Woman is difficult to fathom. You have to be prepared for everything. I could understand her running away from a loving husband after a few months of wedded bliss, but why to Battersea Park! It seemed an inadequate place as a refuge for a disappointed woman.

The jilted husband explains that his wife went there nearly every day, learning to cycle with different instructors. He tried to join her but was too slow and it was not something he enjoyed: 'I followed her for miles. Don't ask me to go again; I can't bear it.' He feared the extent of this widespread attraction. 'This thing', he exclaimed, 'is undermining the femininity of the nation'. No one was safe from the spread of cycling. Even his Aunt Jane, a 'heavily built, elderly lady, of strong pronounced Evangelical views', was at it. She was apparently wearing her Rationals 'in the house to get used to them' but he was certain that 'you will see her riding in them in Hyde Park!'[19]

Figure 10.5 A sunny morning in Battersea Park, *Cycling World Illustrated*, 1896

Given the popularity of Hyde Park cycling, it is perhaps not surprising that Lady Harberton often organised group activities and led rides through the park in rational dress. It was well known that she was not easily intimidated by social scandal but she declared 'that she rode in Hyde Park daily, and met with no unkind or discourteous remarks'. She also said that 'even if the street lad did venture to say hard words concerning her dress, she preferred hard words to broken bones.'[20] These were probably not the kinds of options most cycling women wanted to have to choose between. Many periodicals such as *The Queen*, voiced the opinion that such a radical costume was not necessary for this kind of cycling:

> In that vast throng not one single woman was riding in rational dress, although one or two had certainly donned well-cut divided skirts – so well cut, in fact, that one hardly noticed that they were divided at all. There is no doubt about it, a well-cut ordinary bicycling skirt is all that is necessary for Park riding, or for any ordinary riding, for the matter of that.[21]

The parks were important places where many kinds of cycling, cyclists and costumes were on display and the parameters of acceptable women's cycling were being configured and re-configured on a daily basis. The fact that Mary's design is called 'The Hyde Park Safety Skirt' tells us that it was an attempt to find a happy medium in the increasingly divisive debate. It was still a skirt, but it was designed for the popular Safety bicycle and worn in a location known for its fashionable set. Her invention seemed to address many of the issues. It inhabited a place between the binaries of ordinary and rational dress. Garments like this equipped wearers for multiple mobilities.

Interviewing Mary's Convertible 'Hyde Park Safety Skirt'

Mary outlines the problems of conventional dress that she seeks to address with her patent. 'The danger of ladies wearing an ordinary walking dress when riding a cycle being so great through the liability of being caught in the machine.' She starts, like many before her, with an ordinary full-length A-line skirt. However, this version differs in that it is made up of two separate pieces – front and rear panels – joined together at the waistband. Mary calls it an 'apron'. The two pieces overlap at the sides and are affixed with corresponding buttons and buttonholes. The convertible system operates via two straps attached to the waistband with the same number of buttonholes (seven) as the skirt has buttons. However the straps are shorter than the skirt, so the buttonholes are more closely spaced. The straps turn into gathering devices when they catch each skirt button, or alternatively every second or third button. This method shortens the skirt to a desired length. Mary explains:

> The two edges of each division are then buttoned one over the other; and the two straps having holes worked down them as convenience spaces apart are also buttoned over the divisions; and by missing one or two or more buttons when fastening the straps are consequently made to lift and hold the skirt to any height required for the safety of the rider.

While the mechanism for convertibility seems more straightforward in text than others in the collection, there was no illustration in the patent, so we made a toile for this garment before embarking on the final piece. The process of mocking-up the design enabled us to work out the most appropriate fabric weight and locations for the buttons and buttonholes. We also explored the gathering mechanism and settled on the dual strap device attached to the waistband. Mary offered alternatives. She suggested that 'more than two straps may be used if preferred', the use of studs or links in place of buttons and that 'straps may be attached to a separate waistband instead of directly to the skirt'. This latter variation suggests that the device could be added or adapted to other skirts with buttoned edges.

This is an effective system. The strap-button device hoists material up from the ground and out of the way of the wheels and the ruching effect is pleasantly aesthetic. The material 'festoons', much like Alice's pulley system skirt. In this case the drape runs across the front and back of the skirt, rather than over the hips. The wearer has the choice of which buttons and buttonholes to catch with the straps, and this adapts the height of the skirt to whatever length is appropriate for the intended activity. In the context of fashionable high-society park cycling, the wearer could still give the appearance of wearing a modest ordinary skirt and yet raise the material enough out of the wheels to ensure safe cycling.

There are further differences between this and other convertible costumes in the collection. Mary's straps are expressed design features. When in use they sit outside of the skirt. All the other inventions are

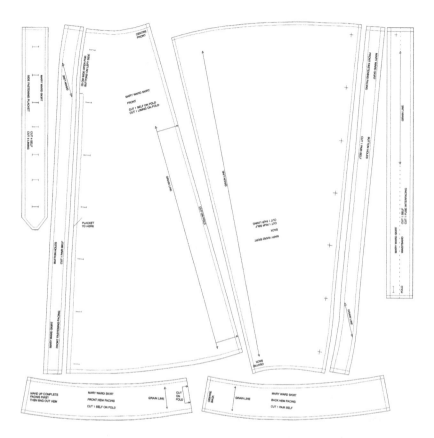

Figure 10.6 The pattern inspired by Mary's convertible cycling skirt consists of eight pieces: front, back, front side facing, back side facing, hem front and back, side strap and waistband

inside; sewn in seams, hidden under peplums or sited under flounces. Another difference lies in the quick-change nature of the invention. For example, Alice's pulley skirt with its weighted hem is very easy to operate – just pull on the cords or release them to reverse the action. Yet even with practice it takes some time to affix Mary's straps to the side skirt buttons. While *The Rational Dress Gazette* article suggested that the skirt 'can be adjusted on the machine in a moment', this seems unlikely. In reality it is fiddly. Even with practice, you would still need two hands. Both straps need to catch buttons at the same intervals to achieve a matching sweep of material. The process of doing this is not difficult, however it seems unlikely that you could cycle and button at the same time. It is more likely that the wearer could undo the buttons while cycling, to drop the skirt in place before jumping off the machine. Given the highly public location of intended use, a key benefit of this design actually lies in its lack of dramatic transformable change. The slowness of conversion, via the button strap system, would have minimised the risk of accidentally revealing undergarments or limbs.

Mary argued that her design was useful for walking as well as cycling. She identified one of the usual problems for perambulate women as having to use one or both hands to lift skirts out of the way of water and mud or clear of kerbs and carriages. This made other activities difficult, such as carrying goods or caring for children. She explains: 'The action of the straps lifting the skirt is in reality imitating the way in which a lady gathers and holds up with her hands, her skirt when wishing to keep the bottom part out of danger of being soiled – say from a muddy road – whilst walking.' Mary's strap system eliminated this problem.

Some might see parallels between Mary's invention and what was commonly known as a skirt lifter or skirt holder. These were popular devices worn attached to a waistband or on chatelaines that gripped the lower part of the skirt and suspended it above the ground. Often made of metal, they were durable and portable; wearers moved them from skirt to skirt. The unsafe, unhygienic and generally annoying problem of women's skirts dragging along the ground was motivating to other patentees who attempted, like Mary, to build these advantages *into* the skirt itself.

Florence Donnelly, a gentlewoman of Manchester, registered a patent in 1895 for an 'Improved Skirt-Lifter and Suspender.'[22] This appliance was located inside the back of the skirt and used a combination of cloth, eyelets, rivets and leather to hoist the material up and out of the way of the ground. A completely different version was patented by Marie Augensen, a manufacturer and spinster, in 1897.[23] Her device featured a length of chain or cord attached to the outside back of the skirt that ran horizontally around the body. The wearer pulled on two ends located at the sides and the skirt was lifted up off the ground. Mary's design is unique in that she tailors it around the needs of the lady cyclist.

Figure 10.7 Converting Mary's cycling skirt

Table 10.1 Requirements

Skirt materials	Extra garments
2.5m Dashing Tweeds – Raver tweed	Waistcoat – 1m Dashing Tweeds New
3m cord trim and bias	Turquoise
50cm printed silk lining for straps	Blouse
14 buttons	Bloomers
	Boots, tights, scarf, hat

Designing a skirt lifter *into* a cycling costume had several advantages over detachable devices. First, it could not become unattached or worse, snagged on things. Mary's straps remained safely hidden inside the skirt until activated and then lay flat against the apron's buttoned edges. Second, her device was made of fabric so it was light and relatively easy to add to garments.

What emerges in Mary's patent and through related devices is a central problem that plagued mobile women and how inventors sought to remedy it. Ordinary skirts were long and streets were muddy and busy. Lifting them out of dirt and danger made a woman even less mobile. Skirts complicated everyday forms of mobility such as walking and accessing vehicles. It was even worse on a bicycle. The many open moving parts of a velocipede amplified these difficulties. More specifically, Mary's design provides insight into how some designers were seeking solutions for a specific kind of urban mobility – park cycling. This was very different to cycling in other public spaces or outside the city. Here, cyclists were on display. They were scrutinised and judged. There was apparently 'hardly a better place ... for seeing the best style of fashions worn in London.'[24] What you wore to cycle reflected who you were and also who you wanted to be. Convertible costumes like Mary's opened up the material and discursive possibilities of how women could move in public space.

Part III

Conclusion: The Politics of Patenting (or How to Change the World One Garment at a Time)

Look at what has happened, and consider the present: – Twenty-five years ago (1873) – men arrested and knocked down in the road for riding bicycles. Fifteen years ago – ladies howled at and almost assaulted for riding tricycles. Six years ago – ladies howled at and threatened by mobs for riding bicycles in everyday costume. Present day – ladies insulted for riding bicycles in cycling costume. Consider the above, and it will show how much as been won. The present hostility of the public is only transitory ... The present hostility is a good omen. It is always darkest just before the dawn.

A. Wheeler (S. S. Buckman)
Wayside Jottings, The Lady's Newspaper, 1898[1]

This is not a conventional heroic account of cycling. It's not about winning races and holding trophies aloft to screaming crowds. There are no awards, only patchy recognition and a bit of commercial success, and even then it is not certain. The ability to track and trace women's lives and contributions to historic inventions can only ever be partially known due to the dominance of patriarchal knowledge systems coupled with the patenting practice of rebranding that erases the inventor. This book is also not about best design practice. We cannot fully assess how impactful these ideas were, if they even worked and how the women who made and wore them really felt about them. Some designs travelled around the world, while others did not seem to get much further than the patent office. In many cases the very nature of their inventions

inversely ensured their erasure from history. As we discovered in doing
this project, even with the costume in hand, these designs remained
stubbornly reluctant to reveal themselves. Some do not work well off
the body. They require the labour and commitment of imagination
and embodiment to come to life. But this is also the point. Camouflage
and concealment were the very essence of their enterprise. These garments were deliberately designed to enable and empower women to
move, but only when, where and how they wanted. Yet, it is the case that
convertible cycle wear inventors did their jobs so well, by deliberately
concealing their inventions from view, that history has all but forgotten
them.

What then do these stories do? Why do they matter?

Matthew Sweet in *Inventing the Victorians* asks: 'Suppose that everything we think we know about the Victorians is wrong.'[2] He revisits historical materials with a desire to reconsider how we as moderns narrate
the past. Contemporary societies invariably consider themselves vastly
more advanced in every way to those who have gone before. Sweet wonders if the Victorians were actually more liberal and fun, and deviant
and experimental than we give them credit for. Perhaps we have misread
them. Perhaps the stories we keep telling are not the *only* stories to tell.

Cycling's history seems similarly familiar (and one-dimensional).
Technological trajectories of the bicycle present a story of advancement.
We can map out where have been, why we are where we are now and
where we might be headed. But there are gaps. Cycling's history is written
predominantly as an account of men's relationships with technology, and
where women do appear within this narrative, they are often consigned
to the role of onlookers, their contributions and achievements overlooked or diminished. There are, for instance, very few detailed records
of women riding early-nineteenth-century velocipedes. When they do
appear in the media accounts, it is more often in relation to terrible
crashes and technological incompetency. Those who raced were largely
depicted as novelty acts. Even their impressive endurance feats tend to
be narrated as anomalous events and not as part of a long, connected
and evolving history. Rarely were they represented as 'proper' cyclists,

more commonly narrated as passive or ornamental bystanders; caught up in waves of technical change; as symbols of social upheaval, not catalysts of it.

Casting women in minor roles effectively erases them from historical narratives. Yet, we can be certain that there were many women who participated in cycling's past, and whose stories have been disregarded or undervalued in line with gendered norms of the time. They may have inhabited less than public spaces, as Diana Crane has argued, which rendered them 'invisible', such as working women in rural areas who wore more practical masculine clothing. They may have been successful in avoiding society's judgemental gaze or disguising themselves as male riders in order to take to the wheel. There were definitely more who invented and co-invented things, but their contributions have gone unrecorded or their traces erased before we could follow them. In 1896 American women's right's activist, Susan B. Anthony may have famously said that the bicycle 'has done more to emancipate women than any one thing in the world',[3] but we seem to know remarkably little about what women have done to emancipate the bicycle. Ostensibly, what I am arguing is that just because women are not in the public record doing things does not mean they were not done. We just need better ways of looking for them.

Inspiration for this task can be found in similar examples of spirited women being unearthed from the past. Jo Stanley has written about the history of women pirates in *Bold in Her Breeches*,[4] which shares many surprising commonalities with early women's cycling. Piracy offered women exciting new forms of mobility, and with it power and freedom. These adventurous women left behind the restraints of domesticity and with it the subordinate norms of feminine life. Their choices shocked the establishment; their morality was questioned, their sexuality disputed and their pursuit of independent lives considered perverse. Much of this had to do with their adoption of masculine practices and clothing. Although this role seemed to offer to rescript conventional narratives, women who took to the seas were cast as either sexual predators or sexual victims. Jo Stanley focuses on the many myths associated with female pirates and attempts to get underneath and around these thin, two-dimensional caricatures by asking, 'how do stories of piracy change if we add flesh-and-blood women and delete the violent glamour?'[5]

In a similar vein, Julie Wheelwright's *Amazons and Military Maids* documents the long tradition of cross-dressing women soldiers, and women's activities on and around the battlefield.[6] She sifts through hundreds of years of military records, legal accounts, periodicals and personal correspondence to collect, collate and analyse a remarkable collection of women's contributions to military history. James Barry was one such brilliant Victorian military doctor who forged a medical career in the eighteenth century despite insurmountable gender barriers. Born in 1790 in Cork, Margaret Bulkley took on her late uncle's name and network, and in disguise presented herself as his nephew, to became the first female doctor in Britain, over half a century before women were even allowed to study medicine. Dr Barry apparently lived a life of worldly adventure as a highly respected skilled surgeon, pioneering new medical procedures including the first successful caesarean.[7] She was a vegetarian, a humanitarian who treated rich and poor alike, an excellent dancer and renowned for being foul-mouthed and, probably not unrelated, also a legendary duellist, which got her in and out of scrapes. Her birth identity was exposed only after death and then kept hidden in military records until recently. Dr Barry did not fit gender normative scripts, then or indeed for over a century later. A series of books dedicated to Dr Barry's life do much work to reclaim her remarkable contribution to medicine and stubborn rejection of the barriers to her sex.[8]

What these scholars bring to light are remarkable narratives that challenge the heroic norms of men as creative innovators or victors, and of women as spectators, victims or trophies. This book has similar aims. Up until now, little has been known about the radical inventive capacities of early women cyclists. Although still marginal, the history of rational dress is much better known than the stories of women who used the patenting system to sew their way out of the 'dress problem'. It might have come as a surprise to readers that women were so involved in the inventive technological cultures of the Victorian era and that there were so many creative channels through which they contested restrictions on their freedom of movement.

Patented convertible cycle wear is an exciting example of women's inventive contributions to cycling's past. As these stories attest, many responded to the social, material and technical challenges to their freedom of movement with vivid creativity. They were not passive participants. They actively and directly worked with and around barriers that sought to prevent them from cycling and engaging more broadly in public life. Their designs offered the means for women to move independently, un-chaperoned, safely and at speed, and through patenting forged new paths into social, cultural and economic worlds. What they remind us is that not all inventions are told through loud or triumphal narratives. The very nature of success is re-configured here. A lot of the work is deliberately hidden, in the lining, under the surface, in the seams. These inventors put in an awful lot of work to *not be seen*. It is no wonder that these skirts did not make it into today's museums and galleries. But it raises questions about how we account for technology stories that do not fit conventional narratives: what else have women and others invented that is hidden in plain sight? What else don't we know about?

These stories serve to interrupt cycling's smooth socio-technical histories and accepted tropes. As we have seen, how women took to cycling and what they wore was not a case of wearing rational *or* ordinary dress, of producing reputable *or* disreputable identities, of supporting *or* not supporting women's rights, of being good *or* bad cyclists or performing masculine *or* feminine identities. These binaries are not only limiting, they are also unrepresentative of the reality of Victorian life. It is deceptively easy to look at different societies, and especially at those located in dusty archives, and neatly compartmentalise them. Yet, life then, as now, was rife with contradictions and complexity, in all its vivid, lively and messy experience.

Perhaps even more importantly, binaries blind us to the creativity in between. As this book illustrates, there was an abundance of inventiveness brought to life in cycle wear patents. Inventors may have been personally and politically motivated in many different ways yet they shared a common interest in re-configuring gender roles and carving out space for newly mobile citizens. Through something as marvellously mundane as clothing they were attempting to de-centre the universal experience, which was dominated by a narrowly defined masculine body, and open it up to different bodies and new ways of being in and moving through public space. These inventors were modestly making social, political and cultural changes one (hidden) garment at a time. The fact that these dynamic devices are built *into* women's skirts should ensure they are recognised even more for their inventiveness, not less.

So, why do they *still* matter?

The history of cycling is not just about stories of brave men and bicycles. Women have long been active and enthusiastic velocipedists in spite of the many barriers to entry at every turn. They have also been involved in broader cycling cultures, from designing through to patenting and manufacturing at all levels of society.[9] The presence of certain voices and artefacts and the absence of others can tell us a great deal about power and value. Erasing women from the record, or not writing them in in the first place, makes it easy to overlook their value and contributions. Conversely, it holds that rendering women more visible in public makes them harder to ignore in other places. We are forced to ask different questions: not only about who and what is visible but who and what is not, and why. And, if we remember the past differently, might it change how we think about and inhabit the present? Can we imagine and make different futures?

This book is shaped by what Kate Eichhorn calls a 'reorientation to history'[10] and Clare Hemmings's invitation to 'tell stories differently'.[11] What they encourage us to do is not only question what is *in* the archive but also the archive itself. Writing about British women's sewing, cycling and suffrage provocatively re-casts and re-configures conventional knowledges and understandings of inventive pasts. These stories matter because histories of technology innovation tend to focus on loud voices in big authoritative fields, such as science and engineering. Science and technology theorist, Wiebe Bijker, argues that 'the stories we tell about technology reflect and can also affect our understanding of the place of technology in our lives and our society'.[12] Smaller, less triumphal and more mundane or deliberately hidden technologies can easily slip by unnoticed, distorting our understanding of their practical potential and cultural significance. To avoid this, feminist technoscientist Susan Leigh Star advocates looking beyond the surface, to 'unearth some drama', to do 'some digging' and 'restore some narrative to what appears to be dead lists'.[13] This approach invites us to re-imagine people who have lived before, and to consider the ways in which the gendering of past narratives might be differently textured, complicated or queered. In producing this counter-narrative, these stories add badly needed dimensions and layers with new and different voices, bodies and things, and drama and colour, in the form of convertible costumes, to historical dialogues about cycling.

These stories also matter to how we live now. Contemporary cyclists are still riding many of the same streets as their Victorian counterparts. It is still the case that few of us can name many famous non-white male inventors; women rarely feature in popular visual culture on statues, media or currency.[14] The mass motorisation of the urban landscape may have changed, but the animosity between modalities does not seem diminished by time. Cycling bodies remain popular sites for debates about scarce urban resources, citizenship rights, social etiquette and gender inequalities.

And, what we wear *still* plays a critical role in everyday life. People, and particularly women, are still judged by their dress. It enables and constrains movement – it shapes who has permission to move, how, in what ways and where in public space and how they are treated. Women *still* share many of the concerns cyclists had in the 1890s. Writing in 1895, pioneering suffragist Frances Willard suggested that 'the old fables, myths, and follies associated with the idea of women's incompetence to handle bat and oar, bridle and rein, and at last the cross-bar of the bicycle'[15] were at last being treated with the contempt they deserved, but over a century later it is clear that these fables, myths and follies persist when women are still not seen as equivalent participants in cycling cultures. Look closely at the range of cycle wear in a shop and you will likely see quite a lot of men's wear and smaller ranges for 'Women' and 'Kids'. It is similar online. Women's clothes are often marked for 'Ladies', which tend to differ in terms of colour palettes and technical quality to men's. Similarly, many designs for 'Ladies' bicycles' still hold traces of the Victorian notion that women do not need or want high quality or cutting-edge technologies. These ideas continue to be reflected in cycling's media coverage, sponsorship and visual culture – lean lycra-clad men are commonly shown doing much of the cycling, and especially the challenging, sweaty and risky riding.

While *ladylike* cycling might accurately represent a type of cycling and cyclist, some of the time, overall it is a limiting catch-all for the vast potential of women on bicycles. Women's cycling is so much more than this. It was in Victorian times too, and yet the stereotype of women as not being particularly interested in cycling, as weak or being socially bound to uphold ideas around grace, charm and elegance while moving remains remarkably stubborn.[16] There are of course notable exceptions to this. Over the last few years we have witnessed the growth of women-specific organisations, brands and events that place women at the heart of their designs and communications and individuals who work tirelessly to raise the profile of women in cycling. Yet, on the whole the cycle wear industry and how cycling is represented remains male dominated, and this imbalance spans from everyday cycling through the discipline of professional racing.[17] Unfortunately, one area in which women continue to lead in relation to cycling is in terms of casual sexism and street harassment, and this is often in relation to their clothing.[18]

S. S. Buckman was a fierce and relentless campaigner for women's rights and freedom of movement. His quote in *The Lady's Newspaper* in 1898 is poignant. He thought public opposition to differently dressed mobile women would not last. Society would change. 'The present hostility is a good omen', he wrote. 'It is always darkest just before the dawn.' Things have changed, and cycling is once again undergoing a widespread resurgence in Britain. But much more is needed. Being part of this change relies on becoming aware that people have been resisting and challenging discrimination for a very long time, that inequalities are not cemented in place and that we each have a role to play in shaping the future for the next generation.

Stories about (convertible) cycle wear *still* matter because they symbolise a marginalised people's fight for equality. We are not there yet. But the journey started over a century ago and we are on the way.

British Cycle Wear Patents 1890–1900: (for New or Improvements to Women's Skirts for the Purposes of Cycling)

Date	Inventor/ Invention	Theme
1893	Pat. No. 13442 (GB189313442A) Sidonie Meissner, Spinster, Gross Brudeegasse, Dresden, German Empire. 'Garment for Lady Cyclists' (23 September).	*Convertible Costume*
1895	Pat. No. 7292 (GB189407292) James Cornes, Professional Tailor and Cutter, of 37 Melbourne Street, Leicester. 'Improvements in Breeches, Knickerbockers and analogous Garments for Cycling and Riding Purposes' (16 February).	*Bloomers/ Knickers*
1895	Pat. No. 9452 (GB189509452A) Margaret Albina Grace Jenkins, Gentlewoman 13 St. George's Place, Hyde Park Corner, London. 'New or Improved Cycling Dress for Ladies' (5 October).	*Built-in Bifurcation*
1895	Pat. No. 11,850 (GB189511850A) Abbie B. Galloway, No. 9 Lincoln Street, East Somerville, County of Middlesex, State of Massachusetts, USA. 'An Improved Costume, chiefly designed for Lady Bicyclists' (20 July).	*Convertible Costume*
1896	Pat. No. 23,299 (GB189523299A) James Rossdale, Ladies Tailor, and William Arthur Hooke Manager, Foreman Cutter, to J.R. Dale & Co, of 125 Gloucester Road, South Kensington, London. 'Improvements in Cycling Skirts for Ladies' Use' (15 February).	*Built-in Bifurcation*

Date	Inventor/ Invention	Theme
1895	No. 16,062 (GB189516062A) Ida May Rew, Manufacturer, of 124 West 23rd Street, New York. 'Improvements in Ladies' Wearing Apparel for Athletic Apparel' (28 September).	*Built-in Bifurcation*
1895	Pat. No. 19,191 (GB189519191A) Evelina Susannah Furber, of 118B Cromwell Road, in the County of Middlesex. 'Improvements in Bicycle Skirts' (23 November).	*Convertible Costume*
1895	Pat. No. 19,259 (GB189519259A) Benjamin Altman, Merchant, of 25 Madison Avenue, New York City, United States. 'Improvements in Bicycle Skirts' (7 December).	*Tailored Skirt*
1895	Pat. No. 22,302 (GB189522302) Charles Campbell, Edward Whitehead Stringer, and William D'Aggerston Telford, Mantle Manufacturers, trading as Campbell, Stringer and Telford, of 21A Old Change, London EC. 'Improvements in Ladies' Skirts' (28 December).	*Convertible Costume*
1895	Pat. No. 6794 (GB189406794A) Madame Julia Gill, Court Dressmaker, 56 Haverstock Hill, N.W. 'A Cycling Costume for Ladies' (16 February).	*Convertible Costume*
1895	Pat. No. 17,145 (GB189517145A) Alice Louisa Bygrave, Dressmaker, of 13 Canterbury Road, Brixton, London. 'Improvements in Ladies Cycling Skirts' (6 December).	*Convertible Costume*
1896	Pat. No. 19,258 (GB189519258A) Benjamin Altman, Merchant, of 25 Madison Avenue, New York City, United States. 'Improvements in Bloomer Costumes' (11 January).	*Bloomers/ Knickers*
1896	Pat. No. 23,298 (GB189523298) James Rossdale, Ladies Tailor, trading as J.R. Dale & Co, of 125 Gloucester Road, South Kensington, London. 'Improvements in Cycling Skirt for Ladies' Use' (15 February).	*Built-in Bifurcation*

Date	Inventor/ Invention	Theme
1896	No. 192 (GB189600192A) Peter Nilsson, Ladies Tailor and Habit Maker, of 33 Conduit Street, London W. 'A New or Improved Cycling Habit for Ladies' Wear' (15 February).	*Tailored Skirt*
1896	Pat. No. 3057 (GB189603057A) Charles William Edward Towers, Cabinet Maker, 71 Fanshaw Street, Hoxton, London. N. 'Improvements in Skirt Suspenders' (11 April).	*Convertible Costume*
1896	Pat. No. 9753 (GB189509753A) John Gooch, Outfitter, 67 Brompton Road, London. 'Improvements in Ladies' Skirts or Dresses for Cycling' (18 April).	*Tailored Skirt*
1896	Pat. No. 14,767 (GB189514767A) Frederick James Haworth Hazard, City of Toronto, York, Ontario, Canada. 'Improvements in Bicycle Costumes' (11 July).	*Convertible Costume*
1896	No. 20,943 (GB189520943A) Samuel Muntus Clapham, Tailor's Cutter, of 13 Queen's Road, Bayswater, London W. 'A New or Improved Combined Safety Cycling Skirt and Knickerbockers for Ladies' Wear' (18 July).	*Convertible Costume/ Built-in Bifurcation*
1896	Pat. No. 13,832 (GB189513832A) Mary Elizabeth and Sara Anne Pease, Gentlewomen, Sunnyside, Grove Road, Harrogate, Yorkshire. 'Improves Skirt, available also as a Cape for Lady Cyclists' (11 April).	*Convertible Costume*
1896	Pat. No. 11,855 (GB189511855A) Herbert Luey, of 202 Washington Park, Brooklyn, County of Kings, State of New York, USA. 'Improvements in Bicycling Habits' (11 April).	*Built-in Bifurcation*
1896	Pat. No. 13,575 (GB189513575A) Emma Grimes, of Knapton Hall, North Walsham, in the County of Norfolk. 'A Combined Bicycling and Walking Skirt' (16 May).	*Convertible Costume*
1896	Pat. No. 8766 (GB189608766A) Frances Henrietta Muller (Mueller), Gentlewoman, Maidenhead in the Country of Meads, Berks. 'Improvement in Ladies' Garments for Cycling and Other Purposes' (30 May).	*Convertible Costume*

Date	Inventor/ Invention	Theme
1896	Pat. No. 7044 (GB189607044A) Alfred Samuel Phillips, Tailor, of 58 Regent Street, London. 'Improvements in and relating to Ladies' Skirts' (9 May).	*Convertible Costume*
1896	No. 15,659 (GB189615659) Harry Harrison, Ladies' Tailor, of 89 Corporation Street, Birmingham. 'An Improvement in the Skirt of Ladies' Cycling Habits' (22 May).	*Convertible Costume*
1896	Pat. No. 11,822 (GB189611822A) Babette Polich and Otto Beyer, Millners, of the Firm Aug. Polich, of Schlossgasse 1–3 Leipzig, Germany. 'An Improved Skirt or Garment' (18 July).	*Convertible Costume*
1896	Pat. No. 14,059 (GB189614059A) Richard Thomas Smailes, Manufacturer, of 1 and 2 Moreton Terrace, South Kensington, London. 'Improvements in Skirts for Lady Cyclists' (1 August).	*Tailored Skirt*
1896	Pat. No. 15,146 (GB189615146A) Diana Elizabeth Togwell, Schoolmaster's Wife, Earl of Jersey's School, Middleton Stoney, Bicester, Oxon. 'Improvements in connection with Skirts for the use of Lady Cyclists' (8 August).	*Convertible Costume*
1896	Pat. No. 19,760 (GB189519760A) David Smith Legg, Foreman Tailor, of 79 Lavender Hill, London S.W. 'A New or Improved Cycling Skirt for Ladies' Wear' (12 September).	*Built-in Bifurcation*
1896	Pat. No. 23,552 (GB189523552A) Charles Henry Hart, Ladies' Tailor, of 24 Termius Road, Eastbourne and 23 Sackville Street, London W. 'Cycling Habit Skirt' (19 September).	*Convertible Costume*
1896	No. 18,906 (GB189618906) Frederick Hooper, Tailor, of 460 Fulham Road, Walham Green, in the County of London. 'An Improvement in Lower Garments for female Cyclists or Riders' (24 October).	*Built-in Bifurcation*
1896	Pat. No. 3903 (GB189603903A) Frank Asa Johnson, Gentleman, of R.1027. No, 260 Dearborn Street Chicago, Illinois, United States. 'Improvements in Skirts' (14 November).	*Convertible Costume*

Date	Inventor/ Invention	Theme
1896	Pat. No. 24,145 (GB189624145A) Lila Austin, Widow, of East Grove, Cardiff, Glamorgan. 'Improvements in Ladies' Dresses or Skirts for Walking and Cycle Riding' (12 December).	*Convertible Costume*
1897	Pat. No. 26,391 (GB189626391) James Rossdale, Ladies' Tailor, trading as J.R. Dale & Co, of 125 Gloucester Road, South Kensington, London. 'Improvements in Cycling Skirts for Ladies' Use' (9 January).	*Built-in Bifurcation*
1897	Pat. No. 14,058 (GB189614058A) Lily Sidebotham, wife of George Henry Sidebotham, Draper, of Newport in the County of Salop. 'An Improved Appliance for Keeping Dress Skirts in Position while Cycling' (16 January).	*Device to Attach, Stiffen or Secure Skirt*
1897	Pat. No. 4267 (GB189604267A) William Fletcher, Ladies' Tailor, of No. 10 Princes Street, Hanover Square, London. 'Improvement in Ladies' Cycling Skirts' (30 January).	Built-in Bifurcation
1897	Pat. No.19,987 (GB189619987A) William Parker Brough, Engineer, of 'Springfield' Kettering in the County of Northampton. 'An Improved Ladies' Cycling Skirt or Habit' (30 January).	*Convertible Costume*
1897	No. 29,448 (GB189629448A) Helena Wilson, Costumier, of 76 Regent Street, London W. 'A New or Improved Combined garter and Skirt Distender for Cycling and other Skirts' (6 February).	*Garter*
1897	Pat. No. 7133 (GB189607133A) Martha Redhouse, Court Dressmaker, 16 Hinde Street, Manchester Square, Parish of Marylebone. 'Improvements in Ladies' Cycling Skirts' (13 February).	*Tailored Skirt*
1897	Pat. No. 8778 (GB189708778A) William Howard Swingler, Ladies' Tailor, and Henry van Hooydonck, Ladies' Tailor's Cutters, of 15 Hotel Street, Leicester. 'Improvements in Ladies' Costumes adaptable for either Cycling, Riding, or Walking Purposes' (26 February).	*Built-in Bifurcation*

Date	Inventor/ Invention	Theme
1897	Pat. No. 780 (GB 189700780A) Sebastian James Sellick, Tailor and Outfitter, of 23 High Street, Weston-super-Mare. 'Improvements in Ladies' Skirts for Cycling and Ordinary Ware' (27 February).	*Built-in Bifurcation*
1897	Pat. No. 1358 (GB189701358) Charles Bristow, Settler, of 54 Lambton Quay, in the City of Wellington, in the Colony of New Zealand. 'A New or Improved Skirt Attachment for Use by Lady Cyclists' (13 March).	*Device to Attach, Stiffen or Secure Skirt*
1897	Pat. No. 9605 (GB189609605) Mary Ann Ward, of 92 Thomas Street, in the County of Gloucestershire. 'Improvement in Ladies' Skirts for Cycling' (27 March).	*Convertible Costume*
1897	Pat. No. 10,332 (GB189610332A) Emily Christabel Woolmer, Spinster, of The Vicarage, Sidcup, in the County of Kent. 'An Improved Skirt Holder for Lady Cyclists' (17 April).	*Device to Attach, Stiffen or Secure Skirt*
1897	Pat. No. 12,684 (GB189712684A) William Howard Swingler, Ladies' Tailor, and Henry van Hooydonck, Ladies' Tailor's Cutters, of 15 Hotel Street, Leicester. 'Improvements in or relating to Ladies' Cycling Skirts' (7 May).	*Device to Attach, Stiffen or Secure Skirt*
1897	No. 17,920 (GB189617920A) Thomas Henry Brown, Commission Agent, of 12 Lever Street, Manchester. 'Improved Skirt or Garment for Ladies' (12 June).	*Built-in Bifurcation*
1897	Pat. No. 11,422 (GB189711422A) Alexander McKinlay, 1 Lancashire Buildings, Water Street, Manchester in the County of Lancashire. 'Improvements in Ladies' Cycling Dress Protectors' (12 June).	*Device to Attach, Stiffen or Secure Skirt*
1897	Pat. No. 11,941 (GB189711941A) Martha Kate Ross White, of 20 Wellesley Road, Croydon. 'Improvements in Ladies' Skirts, especially intended for Cyclists' (19 June).	*Convertible Costume*

Date	Inventor/ Invention	Theme
1897	Pat. No. 18,327 (GB189618327A) Alfred Roberts, Tube Manufacturer, of 146 Tennant Street Birmingahm, in the County of Warwick, 'An Improved Adjustable Dress Attachment for Ladies/ Use When Cycling' (19 June).	*Device to Attach, Stiffen or Secure Skirt*
1897	Pat. No. 15,323 (GB189615323A) Georgina Roberston, Lecturer on Hygiene and Physiology to National Health Society, London, of 8 Upperton Road, Guildford, Surrey. 'Improvements in Skirts for Cyclists and Others' (3 July).	*Built-in Bifurcation*
1897	Pat. No. 13,691 (GB189713691A) Carri Gibbs Tresillian, Governess, of Marmion Road, Southsea Hants. 'Improvements in or relating to Skirts for use when Cycling or Walking' (17 July).	*Built-in Bifurcation*
1897	Pat. No.9251 (GB189709251A) John Sibald, Buyer, of Great George Street, Hillhead, Glasgow. 'Improvements in Cycling Skirts' (24 July).	*Tailored Skirt*
1897	Pat. No.17,115 (GB189717115A) Marie Clementine Michelle Baudéan, Composing Pianist, of 38 Rue du Chateau d'Eau, Paris, France. 'Improved Knickerbockers Seat with a Movable Side for Female Cyclists, Horsewomen, Huntswomen, and the Like' (21 August).	*Bloomers/ Knickers*
1897	Pat. No. 17,235 (GB189717235A) Charles Josiah Ross, Outfitter, trading as J & G. Ross, of 227 High Street Exeter, County of Devon. 'Improvements of Ladies' Cycling Skirts' (21 August).	*Bloomers/ Knickers*
1897	Pat. No. 23377 (GB1897623377A) Blanche Emily Thompson, Dispenser and Registrar, of the Hospital for Women, Upper Priory, in the City of Birmingham, 'Improved Appliance for Holding Skirts Down When Cycling and the Like' (21 August).	*Device to Attach, Stiffen or Secure Skirt*

Date	Inventor/ Invention	Theme
1897	Pat. No. 1377 (GB189701377A) George Albert Shipman of 26 Filey Street, in the City of Sheffield, Manufacturer and Charles Christopher Walker, Engineer, of The Wicker, Sheffield. 'Improved Appliance for Holding the Skirts for Lady Cyclists' (20 November).	*Device to Attach, Stiffen or Secure Skirt*
1897	Pat. No. 2363 (GB189702363) Reuben Payne, Clothier, of 2 Chichester Street, Belfast. 'A Combined Cycling and Walking Skirt for Ladies' (27 November).	*Convertible Costume*
1898	Pat. No. 5831 (GB189805831A) Esther Matthews, Married Lady, of Abbey Foregate, Shrewsbury and Catherine Carter, Spinster, of Brampton House, Havelock Road, Bell Vue, Shrewsbury. 'Improved Means of Retaining Ladies' Skirts in Position when Cycling' (30 April).	*Device to Attach Stiffen or Secure Skirt*
1898	Pat. No. 10,896 (GB189810896A) Charles Garnett, Solicitor, of 26 North John Street, Liverpool, County of Lancaster. 'A New or Improved Appliance for Adjusting the Dress of Ladies in Mounting Cycles' (12 May).	*Convertible Costume*
1898	Pat. No. 16,881 (GB189816881A) Eva Molesworth, Spinster, of Manor House, Bexley, in the County of Kent. 'Improvements in Cycling Skirts'. [Communicated from abroad by Louisa Mary Dennys, Married, of Nungam-baukam Road, Madras, India] (10 June).	*Convertible Costume*
1898	Pat. No. 30,069 (GB189730069A) Edwin Slatter, Outfitter, and George William Richardson, Foreman Tailor, George Slatter, Outfitter, and George William Richardson, Foreman Tailor, of 6 Carlton Nottingham. 'An Improved Cycling and Walking Skirt' (18 June).	*Tailored Skirt*
1898	Pat. No. 22,437 (GB189722437A) Charles Dawson, Gentleman, of 23 Illminster Gardens, Lavender Hill, London, S.W. 'A New or Improved Cycling Skirt Fastener or Adjuster' (25 June).	*Device to Attach, Stiffen or Secure Skirt*

Date	Inventor/ Invention	Theme
1898	Pat. No. 14,592 (GB189814592A) Henrich Ludwig Andreas Timm, Boat Owner, of Eppendorferbaum, 16, Hamburg, 23 Empire of Germany, 'Device for Holding Down Ladies Skirts in Cycling' (6 August).	*Device to Attach, Stiffen or Secure Skirt*
1898	Pat. No. 9427 (GB189809427A) Elizabeth Charlotte Hawkins, Married, of 15 Ashford Road, New Swindon, Wilts. 'An Improved Means for Retaining Ladies' Skirts in Position for Cycling and the Like' (30 July).	*Device to Attach, Stiffen or Secure Skirt*
1898	Pat. No. 14592 (GB189823804A) Mrs Janet Riddle, Lady, of 7 Carlton Road, St. John's London. S.E. 'An Improvement Skirt Holder' (4 November).	*Device to Attach, Stiffen or Secure Skirt*
1898	Pat. No. 17,768 (GB189817768A) Edward Barnes, George Barnes, Arthur Barnes, Frederick Barnes and James Vincent, Ladies' Tailors and Outfitters, all of Old Christchurch Road, Bournemouth, Hampshire. 'An improved Cycling Skirt' (8 October).	*Convertible Costume*
1898	Pat. No. 51 (GB189800051A) Susan Emily Francis, Spinster, 54 Lambton Quay, Wellington, Colony of New Zealand. 'An Improved Cycling Skirt' (28 October).	*Convertible Costume*
1898	Pat. No. 3956 (GB189803956A) Arthur Henry Bull, of 24, Tailor, of Market Place, Derby, 'An Improved Combination Garment for Ladies' Use when Cycling, Riding or the Like' (17 December).	*Bloomers/ Knickers*
1898	Pat. No. 3459 (GB189803459) Henry Albert Harman, Costumier, of 98 Palmerston Road, Southsea. 'A New or Improved Appliance or Attachment for use in Cycling Skirts and the link' (11 February 1899).	*Tailored Skirt*
1899	Pat. No. 23,244 (GB189823244A) George William Fletcher, Ladies' Tailor, of The Square, Barnstaple in the County of Devon. 'Improvements in Ladies' Cycling Skirts and the Like' (17 June).	*Built-in Bifurcation*

Date	Inventor/ Invention	Theme
1899	Pat. No. 26,321 (GB189826321A) Edwin Slatter, Outfitter, George Slatter, Outfitter, and George William Richardson, Foreman Tailor, all of 6 Carlton Nottingham. 'Improvements in Cycling Skirts' (22 July).	*Tailored Skirt*
1899	Pat. No. 14,673 (GB189914673A) William Edward Vallàck and Eliza Jane Vallàck, Court Milliners, of the Downs, Altrincham, in the County of Chester. 'Improvements in Ladies' Skirts for Cycling' (26 August).	*Built-in Bifurcation*
1899	Pat. No. 7085 (GB189907085A) Robert Skelton, Tailor and Outfitter, of 40 and 41 Aungier Street, Dublin Ireland.'Improvements in or relating to Ladies' Cycling Skirts' (21 October).	*Convertible Costume*
1899	Pat. No. 25,346 (GB189825346A) Jean Milne Gower, Married Woman, of Rossmoyne, Wellesley Road, Sutton, in the County of Surrey. 'Improvements in and connected with Ladies' Skirts' (21 October).	*Tailored Skirt*
1899	Pat. No. 23,804 (GB189823804A) Mrs Janet Riddle, Lady, of 7 Carlton Road, St. John's London. S.E. 'An Improvement Skirt Holder' (4 November).	*Device to Attach, Stiffen or Secure Skirt*
1899	Pat. No. 818 (GB190000818A) George Kemp Scruton, Tailor and Outfitter, of 64 New Street, Birmingham, in the County of Warwick. 'An Improved Skirt for Ladies' Use, particularly adapted for Cycling' (24 November).	*Tailored Skirt*
1899	Pat. No.11568 (GB189911568A) Mary Cooke, Spinster, trading as Christie & Co, of No. 81 Baker Street in the Parish of Marylebone, in the County of Middlesex. 'An Improvement in Cycling Skirts for Ladies' Use' (20 January 1900).	*Tailored Skirt*
1899	Pat. No. 6997 (GB189906997A) Harry Harrison, Ladies' Tailor, 89 Corporation Street, Birmingham. 'An Improvement in the Skirt of Ladies' Cycling Habits' (3 February 1900).	*Built-in Bifurcation*

Date	Inventor/ Invention	Theme
1899	Pat. No. 24,136 (GB189924136A) Katie Ryan, Gentlewoman, of 251 Rice Street, St. Paul, Minnesota, United States of America. 'Skirts for Wear when Cycling or in Wet Weather' (10 March 1900).	*Convertible Costume*
1899	Pat. No. 12,370 (GB189912379A) Frederick Charles Cooper, Tailor's Cutter, of "West Leigh", 27 Upperton Road, Eastbourne Sussex. 'Improvements in Cycling Skirts' (7 April 1900).	*Convertible Costume*
1899	Pat. No. 21,883 (GB189821883A). Adelaide Dunbar Baldwin, Wife of James Baldwin, No. 21 Green Park, City of Bath, Esquire. 'Improved Attachment for the Holding-down Loops of Ladies' Riding and Cycling Skirts' (12 August 1899).	*Device to Attach, Stiffen or Secure Skirt*
1900	Pat. No. 12,229 (GB189912229A) Henry Manning Knight, Manufacturer, trading as Henry Knight & Co, of 3 Fell Street, Wood Street EC, 'Skirt Attachment for Cyclists' (9 June).	*Convertible Costume*
1900	Pat. No. 5996 (GB190005996) Mildred Henrietta Edgar Clark, Spinster, of Whitethorn, The Goffs, Eastbourne, in the County of Sussex. 'Improvements in Ladies' Skirts' (23 July).	*Tailored Skirt*
1900	Pat. No. 16,047 (GB190016047A) Cecil Robert Blandford, Master Tailor, of Council Chambers, Corn Street, Newport, Monmouthshire. 'Improvements in Skirts for Women' (23 March 1901).	*Tailored Skirt*
1900	Pat. No. 6929 (GB190006929A) Ethel Eva Minnie Levien, Gentlewoman, of Madowla, St. Kilda Road. Melbourne, Victoria. Australia. 'Improved Women's Cycling Knickers' (19 May).	*Bloomers/ Knickers*
1900	Pat. No. 9701 (GB190009701A) William Hamilton Forsyth, Clothier, 35 Stokes Croft, Bristol. 'An Improvement in Trousers and Knickers for Cycling and Riding applicable also to under Pants or Drawers' (30 June).	*Bloomers/ Knickers*
1900	Pat. No. 1675 (GB190001675A) Judah Burkeman, Tailor, 399 Commercial Road, Landport, Portsmouth, 'Improved Combination, Walking or Riding Ladies' Dress Skirt' (21 July).	*Convertible Costume*

Notes

Introduction

1 I focus on middle- and upper-class women's cycling and patenting experiences because at the time they were at the cutting-edge of fashion, had cultural cachet and money to embrace new technology and leisure time to enjoy it. They were exposed to media and ideas from the 'new worlds' and had the social capital to press against accepted Victorian conventions of how a woman should look and move in public.

2 See: Jungnickel, Kat. (2017) Making Things to Make Sense of Things: DiY as Research and Practice. In Sayer, J. (ed.) *The Routledge Companion to Media Studies & Digital Humanities*, London: Routledge.

3 Download PDF sewing pattern packs at www.bikesandbloomers.com.

4 Tamboukou, Maria. (2016) *Sewing, Fighting and Writing: Radical Practices in Work, Politics and Culture*, London and New York: Rowman & Littlefield, p. 50.

5 Hemmings, Clare. (2011) *Why Stories Matter: The Political Grammar of Feminist Theory*, Durham, NC: Duke University Press, p. 3.

1 'One Wants Nerves of Iron'

1 The Buckman Papers (1894–1900), of Sydney and Maude Buckman relating to the Rational Dress Movement and Cycling for Women. Accessed at the Hull History Centre.

2 *The Sketch*. (1896) Society on Wheels, 11 March, p. 311.

3 Willard, Frances. (1895) *A Wheel Within a Wheel: How I Learned to Ride the Bicycle, with Some Reflections Along the Way*, New York: Flemming.H. Revell Company, p. 13.

4 Crane, Diana. (2000) *Fashion and its Social Agendas: Class, Gender and Identity in Clothing*, Chicago and London: University of Chicago Press, p. 3.

5　Godley, Andrew. (2006) Selling the Sewing Machine Around the World: Singer's International Marketing Strategies, 1850–1920, *Enterprise & Society*, 7 (2): 266–314.

6　Gordon, Sarah A. (2001) 'Any Desired Length': Negotiating Gender through Sports Clothing, 1870–1925. In Scranton, Philip. (ed.) *Beauty and Business: Commerce, Gender and Culture in Modern America*, New York and London: Routledge, pp. 24–51 (p. 34).

7　See: *The Rational Dress Gazette*. (1899) The Fatal Skirt, No. 12, September, p. 46. Also, *The Lady's Own Magazine* (1898) p. 23, which recounts a tale of 'a lighted match thrown from the top of an omnibus, fell beneath the clothing of a lady who was standing on the pavement' which resulted in her garments erupting in 'a mass of flames'. In The Buckman Papers (1894–1900).

8　*The Rational Dress Gazette, Organ of the Rational Dress League*. (1899) A Skirt Responsible for 'Serious Injuries', No. 9, June, p. 86. Accesssed at the University of Hull.

9　The Buckman Papers (1894–1900).

10　*CTC Monthly Gazette*. (1894) The Ladies' Page, February, p. 62.

11　Simpson, Clare. (2001) Respectable Identities: New Zealand Nineteenth-Century 'New Women' – on Bicycles!, *The International Journal of the History of Sport*, 18 (2): 54–77 (p. 54).

12　*Daily Press*. (1899) Pathos of Chaff, 16 September. In The Buckman Papers (1894–1900).

13　Gordon, Sarah A. (2001: 27).

14　*The Lady Cyclist*. (1898) Why Lady Cyclists Should Always Dress Well, March, p. 30. Accessed at the National Cycling Archive, Warwick University.

15　Campbell Warner, Patricia. (2006) *When the Girls Came Out to Play: The Birth of American Sportswear*, Amherst and Boston: University of Massachusetts Press, p. xix

16　The CTC is a cycling advocacy group that still operates in the UK today as Cycling UK.

17　*CTC Monthly Gazette*. (1883) The New Uniform, July, p. 292.

18　*CTC Monthly Gazette*. (1883) The New Uniform, August, p.313

19　*The Queen, the Lady's Newspaper*. (1897) Rational Costume, 11 September, p. 476.

20　*CTC Monthly Gazette*. (1882) Excelsior Tricycle, Advertisement, No. 6, February.

21　Hargreaves, Jennifer. (1994) *Sporting Females: Critical Issues in the History and Sociology of Women's Sports*, London and New York: Routledge, p. 43.

22　*CTC Monthly Gazette*. (1884) April, p. 129.

23　*CTC Monthly Gazette*. (1884) May, p. 168.

24　*CTC Monthly Gazette*. (1883) The Ladies Dress. Important. December, p. 392.

25　*CTC Monthly Gazette*. (1883) The Ladies Dress, June, p. 270.

26　*CTC Monthly Gazette*. (1883) The Ladies Dress, July, p. 293.

27 *CTC Monthly Gazette.* (1894) The New Uniform for Ladies, March, p. 90 – this costume cost £4 and 4s.

28 Crane, Diana. (2000: 5).

29 Crane, Diana. (2000: 1).

30 Campbell Warner, Patricia. (2006: xvii).

31 Tickner, Lisa. (1987) *The Spectacle of Women: Imagery of the Suffrage Campaign 1907–14*, London: Chatto & Windus, p. ix.

32 Zorina Khan, B. (2000) 'Not for Ornament': Patenting Activity by Nineteenth Century Women Inventors, *Journal of Interdisciplinary History*, 31 (2): 159–195 (p. 163).

33 Pat. No. 7292 (GB189729448A). Helena Wilson, Costumier, of 76 Regent Street, London W, 'A New or Improved Combined Garter and Skirt Distender for Cycling and other Skirts' (6 February 1897). Accessed at the European Patent Office Online Database, www.epo.org/index.html.

2 From the Victorian Lady to the Lady Cyclist

1 Vertinsky, Patricia. (1990) *Eternally Wounded Woman: Women, Doctors, and Exercise in the Late Nineteenth Century*, Urbana and Chicago: University of Illinois Press, p. 73.

2 Russett, Cynthia E. (1989) *Sexual Science: The Victorian Construction of Womanhood*, Cambridge, MA: Harvard University Press (cited in Diana Crane 2000: 16).

3 Robinson, J. (1747) *The Art of Governing a Wife; with rules for bachelors. To which is added, an essay against unequal marriages*, London: Gale ECCO.

4 Vertinksy, Patricia. (1990: 48).

5 Showalter, Elaine. (1987) *The Female Malady: Women, Madness and English Culture, 1830–1980*, London: Virago, p. 145.

6 Weir Mitchell, Silas. (1877) *Fat and Blood and How to Make Them*, Philadelphia: J. B. Lipincott & Co, pp. 41–42.

7 Bassuk, Ellen L. (1985) The Rest Cure: Repetition or Resolution of Victorian Women's Conflicts? In Rubin Suleiman, Susan. (ed.) *The Female Body in Western Culture: Contemporary Perspectives*, Cambridge, MA and London: Harvard University Press, pp. 139–151 (p. 139).

8 Cited in ibid., p. 144.

9 Showalter, Elaine. (1987: 146).

10 Ibid., p. 144.

11 The Victorian era is defined by the period of Queen Victoria's reign which ran from 20 June 1837 to 22 January 1901.

12 Bassuk, Ellen L. (1985: 145).

13 Crane, Diana. (2000: 112).

14 Veblen, Thorstein. (1934) *The Theory of the Leisure Class*, London, p. 181.

15 Showalter, Elaine. (1987: 73).

16 Holcombe, Lee. (1973) *Victorian Ladies at Work: Middle-Class Working Women in England and Wales 1850-1914*, Connecticut: Archon Books, p. 4.

17 Crane, Diana. (2000: 16 and 112).

18 Ibid., p. 118.

19 Dolan, Brian. (2001) *Ladies of the Great Tour*, London: HarperCollins, p. 3.

20 Cresswell, Tim. (1999) Embodiments, Power and the Politics of Mobility: The Case of Female Tramps and Hobos, *Transactions of the Institute of British Geographers*, New Series, 24 (2): 175-192 (p. 179).

21 Holcome, Lee. (1973: 8).

22 *The Hub*. (1896) Danger in Excess: Most Women Try to Ride Too Long and Too Hard, 19 September, p. 207.

23 *The Hub*. (1896) The Tables are Now Turned, 19 September, p. 255.

24 Gordon, Sarah A. (2001) 'Any Desired Length': Negotiating Gender through Sports Clothing, 1870-1925. In Scranton, Philip. (ed.) *Beauty and Business: Commerce, Gender and Culture in Modern America*, New York and London: Routledge, pp. 24-51 (p. 24).

25 Parker, Claire. (2010) Swimming: The Ideal 'Sport' for Nineteenth-Century British Women, *The International Journal of the History of Sport*, 27 (4): 675-689 (p. 683).

26 Campbell Warner, Patricia. (2006) *When the Girls Came Out to Play: The Birth of American Sportswear*, Amherst and Boston: University of Massachusetts Press, p. 68.

27 *The Hub*. (1896) Should Women Cycle: Medical View of the Question, 19 September, p. 269.

28 *The Hub*. (1896) Danger in Excess: Most Women Try to Ride Too Long and Too Hard, 19 September, p. 207.

29 *Bicycling News and Tricycling Gazette*. (1893) Women Awheel, 23 September, p. 1.

30 Hargreaves, Jennifer. (2001) The Victorian Cult of the Family and the Early Years of Female Sport. In Scraton, Sheila and Flintoff, Anne. (eds) *Gender and Sport: A Reader*, London and New York: Routledge, pp. 53-65 (p. 58).

31 Editor of the *Daily Telegraph*, reported in *Bicycling News*. (1893) The Ladies Page, 25 November, p. 7.

32 *The Lady's Own Magazine*. (1898) A Clipping from an American Paper, June, p. 173.

33 Simpson, Clare. (2001: 56).

34 Hanlon, Sheila. (2011) The Way to Wareham: Lady Cyclists in Punch Magazine Cartoons, 1890s. Research website, posted 7 April. Accessed at: www.sheila-hanlon.com/.

35 Constanzo, Marilyn. (2010) 'One can't shake off the women': Images of Sport and Gender in *Punch*, 1901-10, *The International Journal of the History of Sport*, 19 (1): 31-56 (p. 35).

36 *The Lady Cyclist*. (1896) The New Woman, 22 August, p. 430.

37 *The Lady Cyclist*. (1896) Editorial, 31 October.

38 See: *The Lady Cyclist*. (1896) Repaired Her Punctured Dunlop Tyre in 5 Minutes, March, p. 43.

39 *Bicycling News and Sport and Play*. (1895) Lines for Ladies by Marguerite, 28 May, p. 34, quoting from 'The Compromises of Cycling' in *Hearth and Home*.

40 See, for more, Longhurst, Robyn. (2001) *Bodies: Exploring Fluid Boundaries*, London: Routledge.

41 *Lady Cyclist*. (1896) No Wonder! 22 August, p. 419.

42 See: Huang, Jinya. (2007) Queen of the Road: Bicycling, Femininity, and the Lady Cyclist, *Cycle History*, 17: 69–76. Also Smethurst, Paul. (2015) *The Bicycle: Towards a Global History*, London: Palgrave Macmillan.

43 Smethurst, Paul. (2015: 95).

44 *The Lady's Own Magazine*. (1898) Readily Understood, August, p. 51.

45 Simpson, Clare. (2001) Respectable Identities: New Zealand Nineteenth-Century 'New Women' – on Bicycles!, *The International Journal of the History of Sport*, 18 (2): 54–77 (p. 54).

46 *Bicycling News and Tricycling Gazette*. (1893) Women Awheel, 23 September, p. 1.

47 *Bicycling News*. (1893) The Ladies Page, 23 September, p. 198.

48 Willett Cunnington, Cecil. (2013) *English Women's Clothing in the Nineteenth Century: A Comprehensive Guide*, Dover Publications, p. 1.

49 Bicycle portraits change over time to represent a larger range of women cyclists. More discussion of this is in Chapter 9.

50 News cuttings accessed in The Buckman Papers (1894–1900).

51 *The Rational Dress Gazette, Organ of the Rational Dress League*. (1899) The Womanly and Graceful Skirt, 11, August, p. 19.

52 *The Lady's Own Magazine*. (1898) Wayside Jottings by A. Wheeler, October, p. 113. A note declaring that contributions by S. S. B to the *Lady's Own Magazine* were made under the pseudonym A. Wheeler is in The Buckman Papers (1894–1900).

53 Crane, Diana. (2000: 122).

54 *Bicycling News and Sport and Play*. (1895) 24 September, p. 10.

55 *Bicycling News*. (1893) Ladies Page, 9 September, p. 168.

56 *The Rational Dress Gazette, Organ of the Rational Dress League*. (1899) May, p. 29.

57 *Bicycling News*. (1893) Dress Reform in Peril, 23 December, p. 1.

58 *CTC Monthly Gazette*. (1898), March, p. 10.

59 *Daily Mail*. (1898) Letter to the editor by P. Maurice, 12 May. In The Buckman Papers (1894–1900).

60 See: Supplement to the *Rational Dress Gazette, Organ of the Rational Dress League*. (1899) April, No. 7, for a roundup of the broadsheet news coverage of the court case.

61 Hanlon, Sheila. (2011) Lady Cyclist Effigy at the Cambridge University Protest, 1897. Sheila Hanlon website, posted 16 April, accessed at www.sheilahanlon.com.

3 Inventing Solutions to the 'Dress Problem'

1 *Bicycling News*. (1893) Dress Reform For Women, by Ada Earland, 18 November, p. 6.

2 *Bicycling News and Sport and Play*. (1895) Lines for Ladies by Marguerite, 14 May, p. 33.

3 *The Queen, the Lady's Newspaper*. (1895) The Weight of Ladies' Bicycles, 13 July, p. 85.

4 Ibid.

5 See for instance: McLachlan, Fiona. (2016) Gender Politics, the Olympic Games and Road Cycling: A Case for Critical History, *The International Journal of the History of Sport*, 33 (4): 469–483.

6 *The Hub*. (1896) Women More Careful, 22 August, p. 104.

7 Manufacturers did make custom-sized machines for special order but these cost more.

8 *Bicycling News and Sport and Play*. (1895 ibid.).

9 *The Queen, the Lady's Newspaper*. (1896) Cycling, 16 May, p. 886.

10 *Lady's Own Magazine*. (1898) Wayside Jottings, by A. Wheeler (pseudonym of S. S. Buckman), March, p. 83.

11 *Harper's Bazaar*. (1897) Outdoor Woman, 18 September, pp. 786–787 (cited in Campbell Warner, Patricia. (2006) *When The Girls Came Out to Play: The Birth of American Sportswear*, Amherst and Boston: University of Massachusetts Press, p. 122).

12 *The Lady Cyclist*. (1896) Little Essay's by Cynthia, On Dress-Guards, 24 October, p. 719.

13 *Bicycling News and Sport and Play*. (1895) Lines for Ladies by Marguerite, 4 June, p. 33.

14 *Bicycling News*. (1893) The Ladies Page, 18 November, p. 2.

15 *New York Times*. (1897) Women and Cycling by Ida Trafford-Bell, *New York Times* Supplement, Sunday 7 February, p. 11.

16 *New York Times*. (1897) Woman and the Bicycle: Mrs Hudders's Dress and Wheel, 11 October, p. 4.

17 *New York Times*. (1894) The Wheelmen of the Day; How Rapidly They Are Multiplying and a Word of Their Dress, 12 August, p. 18.

18 *Lady's Own Magazine*. (1898 ibid.).

19 *Lady's Own Magazine*. (1897) Wayside Jottings, by A. Wheeler (pseudonym of S. S. Buckman), October, p. 119.

20 Ibid.

21 *The Cycling World Illustrated*. (1896) Array Yourself Becomingly, 24 June, p. 351.

22 The Buckman Papers (1894–1900). See also Jungnickel, Kat. (2015) 'One needs to be very brave to stand all that': Cycling, Rational Dress and the Struggle

for Citizenship in Late Nineteenth Century Britain, *Geoforum, Special Issue: Geographies of Citizenship and Everyday (Im)Mobility*, 64: 362–371.

23 Letter from Kitty to Maude, 13 September 1897, The Buckman Papers (1894–1900).

24 Letter from Kitty to Uriah, 18 July 1897, The Buckman Papers (1894–1900).

25 Ibid.

26 *The Rational Dress Gazette, Organ of the Rational Dress League*. (1899) Correspondence – Irene Marshall, No. 10, July, p. 40.

27 Letter from Lady Harberton to S. S. Buckman, 18 April, 1898, in The Buckman Papers (1894–1900).

28 *The Rational Dress Gazette, Organ of the Rational Dress League*. (1898) Editorial, No. 1, June, p. 2.

29 Letter from Lady Harberton to S. S. Buckman, ibid.

30 See Clarsen, Georgina. (2011) *Eat My Dust: Early Women Motorists*, JHU Press.

31 *Bicycling News and Sport and Play*. (1985) Lines for Ladies by Marguerite, 16 April, p. 18.

32 Letter from Jane to Uriah, 4 August 1897, The Buckman Papers (1894–1900).

33 *The Rational Dress Gazette, Organ of the Rational Dress League*. (1898) The Rational Dress League, June, No. 1, p. 4.

34 *Daily Telegraph*. (1896) Cyclists' Association – Skirts V. Rationals, 28 November. In The Buckman Papers (1894–1900).

35 *The Lady Cyclist*. (1896) Lady Cyclists at Home – Mrs Selwyn F. Edge, March, p. 27.

36 Campbell Warner, Patricia. (2006: 8).

37 *The Lady's Own Magazine*. (1898) The Modern Matchmaker, June, p. 173.

38 Robins Pennell, Elizabeth. (1897) Around London by Bicycle, *Harper's Monthly Magazine*, 95 (June): 489–510.

39 *The Rational Dress Gazette, Organ of the Rational Dress League*. (1899) Correspondence from Madel Sharman Crawford, June, No. 9, p. 36.

40 Bullock Workman, Fanny. (1896) Notes of a Tour in Spain, *The Lady Cyclist*, March, Part 1, Vol. II, March, p. 42.

41 Greville, Violet. (1894) *Ladies in the Field: Sketches of Sport*, London: Ward & Downy, p. 261.

42 Parker, Rosika. (2012) *The Subversive Stitch: Embroidery and the Making of the Feminine*, I. B. Tauris, p. 11.

43 Ibid., p. ix.

44 Gordon, Sarah A. (2007) Sewed Considerable: Home Sewing and the Meanings of Women's Domestic Work. In *'Make it Yourself': Home Sewing, Gender, and Culture 1890–1930*, Columbia University Press, Chapter 1, p. 1.

45 Beaudry, Mary Carolyn. (2006) *Findings: The Material Culture of Needlework and Sewing*, Yale University Press.

46 Warner, Patricia Campbell. (2006: 123).

47 Gordon, Sarah A. (2001) 'Any Desired Length': Negotiating Gender through Sports Clothing. In Scranton, Phillip. (ed.) *Beauty and Business: Commerce, Gender and Culture in Modern America*, London and New York: Routledge, pp. 24–51 (p. 26).

48 Greville, Violet. (1894: 261).

49 *Bicycling News and Sport and Play*. (1985) Lines for Ladies by Marguerite, 7 May, p. 36.

50 *The Lady Cyclist*. (1896) March, p. 38.

51 *Bicycling News*. (1893) Rational Dress for Ladies by Lacy Hiller, G, 30 September, p. 213.

52 *Bicycling News*. (1893) The Costume of the Future, The Ladies Page, 30 September, p. 215.

53 Ibid.

54 *Bicycling News and Sport and Play*. (1895) Lines for Ladies by Marguerite, 28 May, p. 34, quoting from 'The Compromises of Cycling' in *Hearth and Home*.

55 *The Rational Dress Gazette, Organ of the Rational Dress League*. (1898) June, No. 1, p. 2.

56 *Bicycling News and Sport and Play*. (1985) Lines for Ladies by Marguerite, 16 April, p. 18.

57 *The Lady Cyclist*. (1896) A Mistaken Idea, 29 August, p. 451.

4 The 1890s Patenting Boom and the Cycle Craze

1 *The Queen, the Lady's Newspaper*. (1895) Dress Echoes of the Week, 16 November, p. 923.

2 Schwartz-Cowan, Ruth. (1997) Inventors, Entrepreneurs and Engineers. In *A Social History of American Technology*, New York and Oxford: Oxford University Press, pp. 119–148 (p. 120).

3 Ibid., p. 124.

4 The Fourteenth Report of The Comptroller General of Patents, Designs, and Trade Marks, with Appendices for the Year. (1896) Annual Report: Presented to both Houses of Parliament by Command of Her Majesty, London.

5 *Church Weekly*. (1897) Inventions Which Have Made People Rich, 5 February, p. 11.

6 Holcombe, Lee. (1973) *Victorian Ladies at Work: Middle-Class Working Women in England and Wales 1850–1914*, Connecticut: Archon Books, p. 8.

7 *Guardian*. (2010) Divorce Rates Data, 1858 to Now: How Has it Changed? *Guardian News Datablog*, 28 January. Accessed at: www.theguardian.com/news/datablog/2010/jan/28/divorce-rates-marriage-ons.

8 *The American Settler*. (1891) 11 July, p. 2.

9 See: *The Manufacturer and Inventor, London*. (1891) Wages in the Minor Textile Trades, 20 February, p. 64. Wages for women and girls were much

lower than men and boys. For instance, in the textile trades, the maximum wage for a woman was 12s 8d in the field of lacework, and for girls 8s 3d in hosiery (girls earned less for lace, 6s 2d). In comparison, working men could earn more than double.

10 Holcome, Lee. (1973: 68).

11 *The Times*. (1881) Parliamentary Intelligence, House of Commons – Patents for Invention Bill, 15 June, p. 6.

12 Ibid.

13 Swanson, Kara. (2009) The Emergence of the Professional Patent Practitioner, *Technology and Culture*, 50 (3): 519–548 (p. 529).

14 *The Times*. (1881 ibid.).

15 Amram, Fred. (1984) The Innovative Woman, *New Scientist*, 24 May, p. 10.

16 Crawford Munro, J. E. (1884) *The Patents, Designs, and Trade Marks Act, 1883 (46 & 47 Vict. C. 57) with the Rules and Instructions, Together with pleadings, order and precedents*, London, p. xxxvii. Accessed at: https://archive.org/stream/patentsdesignsa01britgoog#page/n6/mode/2up.

17 The First Report of The Comptroller General of Patents, Designs, and Trade Marks, with Appendices for the Year. (1884) Report: January 1st to April 30th, Presented to both Houses of Parliament by Command of Her Majesty, London, p. 15.

18 The Second Report of The Comptroller General of Patents, Designs, and Trade Marks, with Appendices for the Year. (1884) Annual Report: Presented to both Houses of Parliament by Command of Her Majesty, London, p. 2.

19 The Twelfth Report of The Comptroller General of Patents, Designs, and Trade Marks, with Appendices for the Year. (1894) Annual Report: Presented to both Houses of Parliament by Command of Her Majesty, London, p. 3.

20 The Thirteenth Report of The Comptroller General of Patents, Designs, and Trade Marks, with Appendices for the Year. (1895) Annual Report: Presented to both Houses of Parliament by Command of Her Majesty, London, p. 5.

21 The Fourteenth Report of The Comptroller General of Patents, Designs, and Trade Marks, with Appendices for the Year. (1896) Annual Report: Presented to both Houses of Parliament by Command of Her Majesty, London, p. 5.

22 Ibid., p. 3.

23 The Fifteenth Report of The Comptroller General of Patents, Designs, and Trade Marks, with Appendices for the Year. (1897) Annual Report: Presented to both Houses of Parliament by Command of Her Majesty, London, p. 4.

24 Ibid., p. 4. The report stated: 'The accuracy of this presumption can now to some extent be tested, for the last renewal fees have been paid upon the Patents of the year 1884. It will be seen that out of the 9,983 sealed Patents of that years, 451, or 4.5 per cent have been maintained for their full term of 14 years. Now out of the 3,898 sealed Patents of the year 1883 (the last of the old Act) 238, or 6.2 per cent., were maintained for their full term, and speaking generally the percentage of Patents under the old Act that ran their

full course was about 6. It thus appears that the percentage of sealed Patents that run for 1 years has fallen from 6 to 4.5 per cent.; but on the other hand, though the ratio is smaller, the actual number o Patents continued for their full term is considerable larger than before.'

25 Ibid., p. 5.

26 The Sixteenth Report of The Comptroller General of Patents, Designs, and Trade Marks, with Appendices for the Year. (1898) Annual Report: Presented to both Houses of Parliament by Command of Her Majesty, London.

27 Ibid., p. 4.

28 Ibid., p. 5.

29 Ibid., p. 5.

30 *The Times.* (1899) Messrs. Staley, Popplewell and Co. Chartered Patent Agents from 61 Chancery Lane, 6 January, p. 15.

31 Ibid.

32 *The Rational Dress Gazette, Organ of the Rational Dress League.* (1898) The Rational Dress League, June, No. 1., p. 4.

33 *The Dawn: A Journal for Australian Women.* (1896) Answers to Correspondents, 1 November.

34 *The Church Weekly.* (1899) What Women Have Done, 6 October, p. 20.

35 Schwartz-Cowan, Ruth. (1997: 120).

36 Kahn, Zorina B. (1996) Married Women's Property Laws and Female Commercial Activity: Evidence form United States Patent Records, 1790–1895, *The Journal of Economic History*, 56 (2): 356–388 (p. 365).

37 *The Queen, the Lady's Newspaper* (1896) Women as Patentees, 18 January, p. 104.

38 *The Dawn.* (1892) 1 November, p. 19.

39 *Scientific American.* (1870) Female Inventive Talent, 17 September, p. 184 (cited in Helvenston Gray, Sally and Peteu, Michaela, C. (2005) 'Invention, the Angel of the Nineteenth Century': Patents for Women's Cycling Attire in the 1890s, *Dress* 32: 27–42 (p. 37)).

40 Zorina Khan, B. (2005) *The Democratization of Invention: Patents and Copyrights in American Economic Development, 1790-1920*, Cambridge: Cambridge University Press, p. 128.

41 Ibid.

42 Ibid.

43 *The Queen, the Lady's Newspaper* (1896) Women as Patentees, 18 January, p. 104.

44 Wajcman, Judy. (2004) *Technofeminism*, Polity Press, p. 41.

45 Schwartz-Cowan, Ruth. (1983) *More Work for Mother: The Ironies of Household Technology from the Open Hearth to the Microwave*, New York: Basic Books, p. 51

46 Ibid., p. 52.

47 Zorina Khan, B. (2005: 136).

48 The Sixteenth Report of The Comptroller General of Patents, Designs, and Trade Marks, with Appendices for the Year. (1898) Annual Report: Presented to both Houses of Parliament by Command Her Majesty, London, p. 5.

49 Abridgement of Specifications, Class 141, Wearing-Apparel, Period AD (1893-96) Printed by the Patent Office, 25 Southampton Buildings, Chancery Lane, London: Darling & Son, p. vi.

50 *Bicycling News and Sport and Play.* (1895) Patents, 19 November, p. 48.

51 Zorina Khan, B. (2005: 136).

52 See: Lampland, Martha and Star, Susan Leigh. (eds) (2009) *Standards and Their Stories: How Quantifying, Classifying and Formalising Practices Shape Everyday Life*, Ithaca and London: Cornell University Press; Bowker, Geoff and Star, Susan. (2000) *Sorting Things Out: Classification and its Consequences*, Cambridge, MA: MIT Press, p. 3.

53 Bowker, Geoff and Star, Susan. (2000: 3).

54 Eichhorn, Kate. (2013) *The Archival Turn in Feminism: Outrage in Order*, Philadelphia: Temple University Press, p. 2.

55 Ibid., p. 3.

56 Stoler, Ann Laura. (2002) Colonial Archives and the Arts of Governance, *Archival Science*, 2: 87–109 (p. 87).

5 *Extraordinary Cycle Wear Patents*

1 Erskine, F. J. (2014) *Lady Cycling: What to Wear & How to Ride*, first published in 1897 by Walter Scott Ltd, London: The British Library, p. 8.

2 See full list of patents at the end of the book.

3 Crane, Diana (2000: 99).

4 Pat. No. 19,760 (GB189519760A) David Smith, Foreman Tailor, 79 Lavender Hill, London. S.W. 'A New or Improved Cycling Skirt for Ladies' Wear' (12 September 1896). All patents accessed at The European Patent Office Online Database, www.epo.org/index.html.

5 Pat. No. 9251 (GB189709251A) John Sibald, Buyer, 7 Great George Street, Hillhead, in the County of the City of Glasgow. 'Improvements in Cycling Skirts' (24 July 1897).

6 Pat. No. 11,941 (GB189711941A) Martha Kate Rose White, 20 Wellesley Road, Croydon. 'Improvements in Ladies' Skirts, especially intended for Cyclists' (19 June 1897).

7 *The Lady Cyclist.* (1896) Letters of a Lady Cyclist, March, p. 20.

8 Pat. No. 14,058 (GB189614058A) Lily Sidebotham, wife of George Henry Sidebotham, Draper, of Newport in the County of Salop. 'An Improved Appliance for Keeping Dress Skirts in Position while Cycling' (16 January 1897).

9 Pat. No. 10,332 (GB189610332A) Emily Christabel Woolmer, Spinster, of The Vicarage, Sicup. 'An Improved Skirt Holder For Lady Cyclists' (17 April 1897).

10 Pat. No. 11,422 (GB189711422A) Alexander McKinlay, 1 Lancashire Buildings, Water Street, Manchester in the County of Lancashire. 'Improvements in Ladies' Cycling Dress Protectors' (12 June 1897).

11 Pat. No. 192 (GB189600192A) Peter Nilsson, 33 Conduit Street, London. W. Ladies' Tailor and Habit Maker, 'A new or Improved Cycling Habit for Ladies Wear' (15 February 1896).

12 Pat. No. 9753 (GB189509753A) John Gooch, 67 Brompton Road, London, Outfitter. 'Improvements in Ladies' Skirts or Dresses for Cycling' (18 April 1896).

13 *The Lady Cyclist.* (1896) Advertisement: John Gooch – Ladies' Tailor and Cycle Habit Maker, March, back page.

14 Pat. No. 51 (GB189800051A) Susan Emily Francis, Spinster, 54 Lambton Quay in the Colony of New Zealand 'An Improved Cycling Skirt' (29 October 1898).

15 *The Lady Cyclist.* (1896) The Thomas Skirt, March, p. 56..

16 *The Cycling World Illustrated.* (1896) 3 June, p. 269.

17 *The Cycling World Illustrated.* (1896) Chats With Cyclists – A Cyclist's Chaperone: Miss Cave by Mr Bancroft, 24 June, p. 342. Accessed at the National Cycling Archive, Warwick University.

18 *The Hub.* (1896) Should Women Cycle: Medical View of the Question, 19 September, p. 269.

19 Jerome, Jerome K. (2016) Women and Wheels (written in 1897). In Spencer, Charles, Pain, Barry, Jerome, Jerome K., and Herschell George. *Cycling: The Craze of the Hour*, Pushkin Press, pp. 82–91.

20 *The Cycling World Illustrated.* (1896) Cycling Whispers, 25 March, p. 32.

21 Pat. No. 9452 (GB189509452A) Margaret Albinia Grace Jenkins, Gentlewoman, of 13 St. George's Place, Hyde Park, in the County of London. 'New or Improved Cycling Dress for Ladies' (5 October 1895).

22 Pat. No. 17,235 (GB189717235A) Charles Josiah Ross, Outfitter, trading as J & G. Ross, of 227 High Street Exeter, in the County of Devon. 'Improvements in Ladies' Cycling Skirts' (21 August 1897).

23 Pat. No. 20,943 (GB189520943A) Samuel Muntus Clapham, Tailor's Cutter, of 13 Queens Road Bayswater, London. 'A new or Improved Combined Safety Cycling Skirt and Knickerbockers for Ladies' Wear' (18 July 1896).

24 Pat. No. 17,115 (GB189717115A) Marie Clementine Michelle Baudéan, of 38 Rue du Chateau d'Eau, Paris France, Composing Pianist. 'Improved Knickerbockers Seat with a Movable Side for Female Cyclists, Horsewomen, Huntswomen and the like' (21 August 1897).

25 Pat. No. 7292 (GB189407292A) James Cornes, Professional Tailor and Cutter, of 37 Melbourne Street. Leicester, Professional Tailor and Cutter, 'Improvements in Breeches, Knickerbockers and analogous Garments for Cycling and Riding Purposes' (16 February 1895).

26 Pat. No. 19,258 (GB189519258A) Benjamin Altman, Merchant, of 25 Madison Avenue, New York City, United States. 'Improvements in Bloomer Costumes' (7 December 1895).

27 *The Lady Cyclist.* (1896) The Paris Mode – The Jupe-Culotte, 27 June, p. 172.

28 *The Lady Cyclist.* (1896) Lady Cyclists At Home – Mrs Paul Hardy, 22 August, pp. 433–434.

29 *The Rational Dress Gazette, Organ of the Rational Dress League.* (1899) No. 7, April, p. 26.

30 *The Lady Cyclist.* (1896) Cycling Fashions by Madame Mode, 27 June, p. 167.

31 *The Lady Cyclist.* (1896) Cycling Fashions by Madame Mode, 22 August, p. 426.

32 *The Lady Cyclist.* (1896) Dress Well, 26 September, p. 581.

33 This was also not for *all* women. Voting rights at this time were granted to those over 30 years of age and with property to their name. For younger women this was even later, in 1928.

34 See for example; Hammersley, Martyn and Atkinson, Paul. (2007) *Ethnography: Principles in Practice*, Third Edition, Oxon: Routledge; and Skeggs, Bev. (2007) Feminist Ethnography. In Atkinson, P., Coffey, A. J., Delamont, S., Lofland, J., and Lofland, L. H. (eds) *Handbook of Ethnography*, London: Sage, pp. 426–442.

35 Jungnickel, Kat. (2017).

6 Patent No. 17,145: Alice Bygrave

1 Pat. No. 17,145 (GB171451895) Alice Louisa Bygrave, Dressmaker, of 13 Canterbury Road, Brixton, in the County of Surrey. 'Improvements in Ladies' Cycling Skirts' (6 December 1895). Accessed at the European Patent Office Online Database, www.epo.org/index.html.

2 *The Queen, the Lady's Newspaper.* (1896) Women as Patentees, 18 January, p. 104.

3 See for example: *Bicycling News.* (1893) Infringement of a Cyle Lamp Patent – Important Decision, 15 April, Back cover.

4 Office for National Statistics; National Records of Scotland; Northern Ireland Statistics and Research Agency (1871): Census aggregate data. UK Data Service.

5 Wells, H. G. (2016) *Time Machine*, Wisehouse Classics (originally published in 1895).

6 Post Office London Directory, Official, Street, Commercial, Trades, Law, Court, Parliamentary, Postal, City and Clerical, Conveyance and Banking Directories (1895). Available at: http://specialcollections.le.ac.uk/cdm/ref/collection/p16445coll4/id/8842.

7   This would have been an eventful period of time for Alice as her mother, Emma, apparently died at 49 years old in 1882.

8   Office for National Statistics; National Records of Scotland; Northern Ireland Statistics and Research Agency (1891): Census aggregate data. UK Data Service.

9   Incidentally, the Census at this time records the Duerre household as being completely different. Charles was living alone in Chelsea, now with Emma passed away and all the children moved out. The only other person recorded with him on the night was the housekeeper, Annie E. Trummer, 21.

10   *Bicycling News.* (1893) Our Contemporaries, 7 October, p. 228.

11   *Bicycling News.* (1893) Ladies Page, 23 September, p. 198.

12   Pat. No. 18,937 (GB189418937A) Rudolph Charles William Duerre, of 466 Kings Road Chelsea, County of Middlesex, watch and Clock Maker, 'Improvements in Cycle Saddle Springs' (29 December 1894).

13   Pat. No. 11,911 (GB189811911A) Rudolph Charles William Duerre, of 466 Kings Road Chelsea, County of Middlesex, watch and Clock Maker, 'Improvements in Cycle Saddle and Seat Springs' (15 April 1899).
      Pat. No. 980, Rudolph Charles William Duerre, Austrian Patent – 'Federnde Stütze für Fahrradsättel' (26 February 1900)

14   Pat. No. 21,285 (GB190021285A) Rudolph Charles William Duerre, of 466 Kings Road Chelsea, County of Middlesex, Watch and Clock Maker, 'Improvements in Cycle and the like Saddles' (14 September 1901).
      Pat. No. 22,065 (GB190322065A), Pat. No. 11,911 (GB189811911A) Rudolph Charles William Duerre, of 466 Kings Road Chelsea, County of Middlesex, Watch and Clock Maker, 'An Improvement in Saddles for Bicycles and Other Velocipedes' (18 August 1904).

15   35,328 readers accessed library materials between 4 and 10 p.m. in 1897. The Sixteenth Report of The Comptroller General of Patents, Designs, and Trade Marks, with Appendices for the Year. (1898) Annual Report: Presented to both Houses of Parliament by Command Her Majesty, London.

16   Rosina and Arthur were married on 3 July 1883 in Battersea.

17   *CTC Monthly Gazette.* (1886) A Growing Abuse, August, p. 322.

18   Kinsey, Fiona. (2011) Stamina, Speed and Adventure: Australian Women and Competitive Cycling in the 1890s, *The International Journal of the History of Sport*, 28 (10): 1375–1387 (p. 1376).

19   Simpson, Clare. (2016) Capitalising on Curiosity: Women's Professional Cycle Racing in the Late-Nineteenth Century. In Horton, D., Rosen, P. and Cox, P. (eds) *Cycling and Society*, London and New York: Routledge (first published by Ashgate in 2007), pp. 47–66 (p. 52).

20   *Pall Mall Gazette.* (1896) Cycling Ladies at the Aquarium, 10 December. Available at: http://sixday.org.uk/html/1896_london.html.

21   *Sydney Morning Herald.* (1896) 2 December, p.

22 *The Queen, the Lady's Newspaper.* (1895) Ladies' Outdoor Amusements, 7 December, p. 1095.

23 Lilian was born 1884, Winifred in 1888, Alma in 1892 and Gladys in 1894.

24 *The Sketch.* (1896) Society on Cycles, 22 January, p. 686.

25 *Reynold's News, London.* (1896) Royal Aquarium Last Night: Finish of Novel Ladies' Cycle Race.

26 The most miles were cycled by Miss Blackburn (447), followed by Miss Farrar (421) and Miss Pattison (401).

27 Rosina was clearly an important part of the family. The youngest child of the Duerre family, Bertha, was born 2 September 1877 and given the middle name Rosina.

28 *The Times.* (1899) 9 March, p. 14.

29 Many thanks to Nikki Pugh and Alice Bygrave's relatives for alerting me to this part of the story.

30 *The Rational Dress Gazette, Organ of the Rational Dress League.* (1898) Editorial, June, No. 1, p. 2.

31 *The Queen, the Lady's Newspaper.* (1896) The Bygrave 'Quick Change' Cycling Skirt, 4 April, p. 595.

32 *The Wheelwoman and Society Cycling News.* (1896) Jaegar Advert, 28 November, p. 34.

33 *The Lady Cyclist.* (1896) All Wool, March, p. 49.

34 *The Tailor and Cutter.* (1895) Cycling Garments at the Stanley Show, 5 December, p. 485.

35 *The Wheelwoman and Society Cycling News.* (1896, ibid.).

36 For more about Victorian working wages, see: www.waynesthisandthat. com/servantwages.htm.

37 *The Westminster Budget.* (1896) Concerning Hearth and Home: Wear on the Wheel, 3 April, p. 30.

38 *The Saint Paul Daily Globe.* (1896) $5000 in Four Days: Mrs Bygrave received it last week for a bicycle skirt, 15 March, p. 19.

39 *San Francisco Chronicle.* (1896) For the Girl Who Would Go-A-Wheeling, 12 April, p. 8.

40 *The Worker* (Brisbane, Qld). (1896) Sporting. 23 May, p. 10. Accessed at the National Library of Australia http://trove.nla.gov.au.

41 *Sydney Morning Herald.* (1896) 26 September, p. 5.

42 *The Sydney Daily Mail and New South Wales Advertiser.* (1896) The Ladies' Page, 5 December, p. 1200 Accessed at the National Library of Australia http://trove.nla.gov.au

43 *The Australasian.* (1896) Social Notes, 12 December, p. 42. Accessed at the National Library of Australia http://trove.nla.gov.au.

44 *The Australasian.* (1897) Indispensible to Cyclists, Jaeger Advertisement, 16 January, p. 141. Accessed at the National Library of Australia http://trove.nla. gov.au.

45 As Marianne de laet and Annemarie Mol have illustrated in their study of the Zimbabwe Bush Pump 'B', the term 'working' is far from straightforward when technologies continue to operate in ways unintended by the designer. See: de Laet, M. and Mol, A. (2000) The Zimbabwe Bush Pump: Mechanics of a Fluid Technology, *Social Studies of Science*, 30: 225-263.

46 *The Wheelwoman*. (1897) A New Saddle Cover, 23 January, p. 6.

47 Office for National Statistics; National Records of Scotland; Northern Ireland Statistics and Research Agency (1901): Census aggregate data. UK Data Service.

48 Office for National Statistics; National Records of Scotland; Northern Ireland Statistics and Research Agency (1911): Census aggregate data. UK Data Service.

49 The 1901 Census records show Charles still residing in Chelsea, but now with Bertha, who was 23 at the time, not registered as working and single. Ten years later, the 1911 Census indicates that Bertha is still single, working as a domestic servant. Interestingly the woman of the household was a court dressmaker by trade and working as a saleswoman in a Court Costumier. Bertha, like Alice, must have had dressmaker skills and interests. According to the England & Wales, Civil Registration Index 1916-2000, she died at the age of 80 in 1958, apparently unmarried.

50 Helveston Gray, Sally and Peteu, Michaela. (2005) Invention, 'the Angel of the Nineteenth Century': Patents for Women's Cycling Attire in the 1890s, *Dress*, 32: 27-42.

51 Helveston Gray and Peteu also see similarities with Clarissa Ellen Dockham's 'Bicycle- Skirt' Patent No. 568,339 (1896) and Eliza Voorhis's 'Cycling-Skirt', Patent No. 584,909 (1897) (2005: 41). And there were others years before Alice's patent. The *CTC Gazette* published a letter from a Mrs Jessie Kate Powell of Melton Mowbray in 1883. She describes a skirt she adapted for cycling which had 'six sets of rings down the inside of skirt, with cords running through them, three of which cords come through a worked hole in the waistband, at the side and the other three behind. When walking, the strongs are pulled up and looped over a button' (*CTC Monthly Gazette*. (1883) The Ladies Dress, August, p. 320).

52 For more on the idea of 'immutable mobiles' and 'black boxed' information see: Latour, Bruno and Woolgar, Steve. (1979) *Laboratory Life: The Social Construction of Scientific Facts*, London: Sage.

7 Patent No. 6794: Julia Gill

1 Pat. No. 6794 (GB67941894) Madame Julia Gill, Court Dressmaker, of 56 Haverstock Hill, N. W. 'A Cycling Costume for Ladies' (16 February 1895). Accessed at the European Patent Office Online Database: www.epo.org/index.html.

2 Peteu, Michaela Cornelia and Helveston Gray, Sally. (2009) Clothing Invention: Improving the Functionality of Women's Skirts, 1846–1920, *Clothing & Textiles Research Journal*, 27 (1): 45–61 (p. 48).

3 Pascoe, London of Today 903, cited in Evans, Hilary and Evans, Mary. (1976) *The Party that Lasted 100 Days: The Late Victorian Season: A Social Study*, London: Macdonald and Jane's, p. 131.

4 *The London Illustrated.* (1895) London Season, 1 June, p. 15.

5 Evans, Hilary and Evans, Mary. (1976: 3).

6 Ibid., p. 6.

7 Ibid., p. 26.

8 Ashelford, Jane. (1996) *The Art of Dress: Clothes and Society*, London: The National Trust, p. 173.

9 Evans and Evans note an article written in October 1901 in the *Harmsworth London Magazine* called 'The Impossibility of Dressing on £1000 a year' (1976: 27).

10 Ibid., p. 148.

11 *The Tailor and Cutter.* (1896) The Slack Season, Vol. 31, No. 1,532, 13 February, p. 67.

12 *The Queen, the Lady's Newspaper.* (1895) Dress Echoes of the Week, 30 March, p. 560.

13 *The Lady Cyclist.* (1896) Queens and Princesses, March, p. 8.

14 *The Lady Cyclist.* (1896) Editorial by Charles P. Sisley, March.

15 *The Rational Dress Gazette, Organ of the Rational Dress League.* (1899) Tailor made, No. 13, October, p. 52.

16 *The Lady Cyclist.* (1896) Taste in Dress, March, p. 14.

17 *The Lady Cyclist.* (1896) A Novel Skirt, March, p. 46.

18 *The Queen, the Lady's Newspaper.* (1895) 16 November, p. 923.

19 Pat. No. 19,191 (GB18951919A) Evelina Susannah Furber, 118B Cromwell Road, in the County of Middlesex. 'Improvements in Bicycle Skirts' (23 November 1895).

20 *The Queen, the Lady's Newspaper.* (1896) The New Cycling Costumes of the Season, by Various Makers, 11 July, p. 83.

21 Rappaport, Erika Diane. (2000) *Shopping for Pleasure: Women in the Making of London's West End*, New Jersey and West Sussex: Princeton University Press, p. 5.

22 Peiss, Kathy. (2001) On Beauty … and the History of Business. In Scranton, Phillip. (ed.) *Beauty and Business: Commerce, Gender and Culture in Modern America*, New York and London: Routledge, pp. 7–23 (p. 11).

23 Crane, Diana. (2000: 5).

24 Gamber, Wendy. (1997) *The Female Economy: The Millinery and Dressmaking Trades, 1860–1930*, Urbana: University of Illinois Press, p. 10.

25 *The Tailor and Cutter.* (1895) Cycling Garments at the Stanley Show, 5 November, p. 485.

26 *The Ladies' Tailor*. (1897) Illustrations of British Costumes, Spring 1897, March, No. 3, Vol. 13, p. 25. Accessed at The British Library.

27 Ibid., and *The Ladies' Tailor*. (1897) Notes of the Month, Spring 1897, March, No. 3, Vol. 13, p. 26. Accessed at The British Library.

28 *The Ladies' Tailor*. (1898) Illustrations of Ladies' Cycling Costume, March, No. 3, Vol. 14, p. 26. Accessed at The British Library.

29 *The Lady Cyclist*. (1896) PHOTOGRAPHS! The Prettiest Cyclist in the Prettiest Costume! A novel form of prize, 31 October, p. 761.

30 Gordon, Sarah. (2001) 'Any desired length': Negotiating Gender through Sports Clothing, 1870-1925. In Scranton, Philip. (ed.) *Beauty & Business: Commerce, Gender and Culture in Modern America*, New York and London: Routledge, pp. 25–51 (p. 39).

31 *The Lady Cyclist*. (1896) 'En Avant' Cycling Skirt Advertisement, March, p. 46.

32 *The Lady Cyclist*. (1896) Famous Cycle Emporiums, 22 August, p. 438.

33 *The Queen, the Lady's Newspaper*. (1896) Mr Marcus' New Cycling Room, 4 April, p. 598.

34 Rappaport (2000: 119).

35 *Chamber's Journal of Popular Literature, Science and Art*. (1864) The London Shop-Fronts, 15 October, p. 670. Cited in Rappaport (2000: 119).

36 *The Tailor and Cutter*. (1896) Goods From Shop Windows, 9 January, p. 20.

8 Patent No. 8766: Frances Henrietta Müller

1 Pat. No. 8766 (GB87661896A) Frances Henrietta Müller, Gentlewoman, of Meads, Maidenhead, in the County of Berks. 'Improvements in Ladies' Garments for Cycling and Other Purposes' (30 May 1896). Accessed at the European Patent Office Online Database, www.epo.org/index.html.

2 *Woman's Herald: Women's Penny Paper*. (1891) Interview: Miss F. Henrietta Müller, Editor of the *Woman's Herald*, by Clara E. De Moleys. 28 November, p. 916. Accessed at the Women's Library, London School of Economics.

3 1871 UK Business Register records William Müller, Esq of Hillside, Herts as a Magistrate for Middlesex.

4 *Woman's Herald: Women's Penny Paper*. (1891: 915)

5 *Woman's Herald: Women's Penny Paper*. (1888) Vol. 2(1), 3 November, pp. 4–5.

6 Maudsley, Henry. (1894) Sex in the Mind and in Education, *Popular Science Monthly*, Vol. 5, June, p. 200.

7 Ibid., p. 204.

8 Showalter, Elaine. (1987: 123).

9 Ibid.

10 McWilliams-Tullberg, Rita. (1998) *Women at Cambridge*, Cambridge: Cambridge University Press, p. 71.

11 *Woman's Herald: Women's Penny Paper*. (1891: 916).

12 *The Inter Ocean*. (1884) 16 August.

13 *Woman's Herald: Women's Penny Paper*. (1888) No. 1, Vol. 1, 27 October, p. 4.

14 *Woman's Herald: Women's Penny Paper*. (1888) No. 2, Vol. 1, 3 November, p. 2.

15 *Woman's Herald: Women's Penny Paper*. (1888) No. 6, Vol. 1, 1 December, p. 2.

16 *Woman's Herald: Women's Penny Paper*. (1889) 2 March, p. 7.

17 *The Inter Ocean*. (1889) 2 June.

18 *Evening Star* (Washington, District of Columbia). (1904) Miss F. Henrietta Muller. B.A will give three informal talks to men and women in the Assembly Hall at the Shoreham Hotel, 12 May, p. 12.

19 *Evening Star* (Washington, District of Columbia). (1904) Amusements – Two Lectures will be delivered by Miss. F. Henrietta Muller on Free Motherhood or Parthenogenesis, 12 February, p. 16.

20 Crawford, Elizabeth. (2003) *The Women's Suffrage Movement: A Reference Guide 1886–1928*, Routledge.

21 Cambridge Orlando Project: Women's writing in the British Isles from the Beginning to the Present. Henrietta Muller, http://orlando.cambridge.org.

22 Burman, Barbara, and Denbo, Seth. (2007) *Pockets of History: The Secret Life of an Everyday Object*, Museum of Costume Bath.

23 Fennetaux, Ariane. (2008) Women's Pockets and the Construction of Privacy in the Long Eighteenth Century, *Eighteenth Century Fiction*, 20 (3): 307–334 (p. 308).

24 Ibid., p. 315.

25 Burman, Barbara. (2002) Pocketing the Difference: Gender and Pockets in Nineteenth-Century Britain, *Gender & History*, 14 (3): 447–469 (p. 453).

26 Printed postcard photo of a model wearing Madame Brownjohn's tricycling costume (1885). Accessed at the Women's Library, London School of Economics.

27 *Bicycling News and Sport and Play*. (1895) Lines for Ladies by Marguerite, 16 April, p. 18.

28 Pat. No. 3312 (GB189503312A) Amy Hart, Spinster, of 18 Hildreth Street Balhamn in the County of Surrey. 'Improvements in Pocket Protectors' (14 Dec 1895). Accessed at the European Patent Office Online Database, www.epo.org/index.html.

29 Pat. No. 8316 (GB189708316A) Blanche Ward, of 69 Underhill Road, Lordship Lane, London, S.E. 'Improvements in Pocket Protectors Against Pocket-Picking-' (5 Feb 1898). Accessed at the European Patent Office Online Database, www.epo.org/index.html.

30 Perkins Gilman, Charlotte. (1997) If I were a Man. In *The Yellow Wallpaper and Other Stories*, Dover Thrift Editions, p. 58.

31 Ibid.

32 Ibid.

33 *Woman's Herald: Women's Penny Paper*. (1891 ibid.).

34 *The Rational Dress Gazette*. (1899) Rational Underclothing, No. 10, July, p. 38.

9 Patent No: 13,832: Mary and Sarah Pease

1   Kinsey, Fiona. (2012) Reading Photographic Portraits of Australian Women Cyclists in the 1890s: From Costume and Cycle Choices to Constructions of Feminine Identity. In Huggins, Mike and O'Mahony, Mike. (eds) *The Visual in Sport*, London and New York: Routledge, pp. 35–51 (p. 36).

2   Letter from Kitty to Maude, 13 September 1897, in The Buckman Papers (1894–1900).

3   See Oddy, Nicholas. (2010) Rides on My Safety. The Diary of Emily Sophia Coddington. In *Cycle History*, Birmingham: JPMPF, pp. 29–34.

4   *Bicycling News and Sport and Play*. (1895) Lines for Ladies by Marguerite, 18 June, p. 21.

5   *Bicycling News and Sport and Play*. (1895) Lines for Ladies by Marguerite, 16 April, p. 18.

6   *The Wheelwoman*. (1897) When in Doubt – Consult Cynthia, 11 September, p. 19.

7   *The Morning Leader*. (1899) A Rational Cycling Costume – A Costume for the Country, 6 February. In The Buckman Papers (1894–1900).

8   The Buckman Papers (1894–1900).

9   *The Lady's Own Magazine*. (1897) Wayside Jottings, by A. Wheeler (pseudonym of S. S. Buckman), October, p. 119.

10 *The Rational Dress Gazette, Organ of the Rational Dress League*. (1898) No. 1, June, p. 2.

11 *The Rational Dress Gazette, Organ of the Rational Dress League*. (1899) Editorial, No. 14, November, p. 54.

12 *Daily Mail*. (1898) 11 February. In The Buckman Papers (1894–1900).

13 *The Lady Cyclist*. (1896) Lady Snap-shotters, 26 September, p. 580.

14 *The Lady Cyclist*. (1896) PHOTOGRAPHS! The Prettiest Cyclist in the Prettiest Costume! A novel form of prize, 31 October, p. 761.

15 *The Rational Dress Gazette, Organ of the Rational Dress League*. (1898) No. 1, June, p. 2.

16 Kinsey, Fiona. (2012: 46).

17 Tickner, Lisa. (1987) *The Spectacle of Women: Imagery of the Suffrage Campaign 1907–14*, London: Chatto & Windus, p. 37.

18 Kinsey, Fiona. (2012: 36).

19 Pat. No. US549472A. Alice Worthington Winthrop, 'Bicycle Skirt' (5 November 1895). Accessed at: www.google.as/patents/US549472.

20 Helvenston Gray, Sally and Peteu, Michaela C. (2005) 'Invention, the Angel of the Nineteenth Century': Patents for Women's Cycling Attire in the 1890s, *Dress*, 32: 27–42 (p. 33).

21 Ibid.

10 Patent No. 9605: Mary Ward

1 Lady Harberton to S. S. Buckman, 18 April 1898. In The Buckman Papers (1894–1900).

2 *The Lady's Own Magazine*. (1897) The 'Hyde Park' Patent Safety Skirt, December, pp. 191–192. In The Buckman Papers (1894–1900).

3 Sadly, high quality images were not available for use in this book.

4 *The Rational Dress Gazette, Organ of the Rational Dress League*. (1899) Notes and Comments, No. 7, April, p. 1.

5 *The Queen, the Lady's Newspaper*. (1896) The Culture of the Cycle, 25 January, p. 169.

6 Erskine, F. J. (2014) *Lady Cycling: What to Wear & How to Ride*, first published in 1897 by Walter Scott Ltd, London: The British Library, p. 11.

7 *The Queen, the Lady's Newspaper*. (1896) The Culture of the Cycle, 25 January, p. 169.

8 *The Queen, the Lady's Newspaper*. (1896) Bicycling in Hyde Park, 8 February, p. 258.

9 *The Hub*. (1896) Pars from the Parks, 15 August, p. 67.

10 *The Queen, the Lady's Newspaper*. (1896) Cycling News, 21 March, p. 515.

11 Ibid.

12 *The Queen, the Lady's Newspaper*. (1896, 25 January ibid.).

13 *The Lady Cyclist*. (1896) 17 June, p. 163.

14 *The Queen, the Lady's Newspaper*. (1896, 8 February ibid.).

15 *The Queen, the Lady's Newspaper*. (1896, 25 January ibid.).

16 Atkins, P. (1990) The Spatial Configuration of Class Solidarity in London's West End 1792–1939, *Urban History Yearbook*, 17: 35–65 (p. 56).

17 *Tailor and Cutter*. (1895) Fashions in the Park, 2 May, p. 174.

18 Jerome, Jerome K. (2016: 84–86).

19 Ibid., p. 87.

20 *The Queen, the Lady's Newspaper*. (1896, 25 January ibid.).

21 *The Queen, the Lady's Newspaper*. (1896, 8 February ibid.).

22 Pat. No. 3436 (GB189503436A) Florence Donnelly, Gentlewoman, of 69 Tivoli Place, Higher Broughton, Manchester, Lancashire. 'Improved Skirt-Lifter and Suspender' (21 December 1895).

23 Pat. No. 20,350 (GB189720350A) Marie Augensen, Manufacturer and Spinster, of No. 1606 N. Troy Street, Chicago, Illinois, USA. 'Improved Skirt Raiser and Protector' (5 October 1897).

24 *The Queen, the Lady's Newspaper.* (1896) Hyde Park on Sunday, 18 January, p. 89.

Conclusion

1 *The Lady's Newspaper.* (1895) Wayside Jottings, by A. Wheeler (pseudonym of S. S. Buckman), July, p. 177.

2 Sweet, Matthew. (2001) *Inventing the Victorians*, London: Faber & Faber.

3 *New York Sunday World.* (1896) Champion of Her Sex: Miss Susan B. Anthony Tells the Story of Her Remarkable Life to Nellie Bly, 2 February, p. 10.

4 Stanley, Jo. (1995) *Bold in Her Breeches: Women Pirates Across the Ages*, Hammersmith: HarperCollins.

5 Ibid., p. 15.

6 Wheelwright, Julie. (1989) *Amazons and Military Maids: Women Who Dressed as Men in Pursuit of Life, Liberty and Happiness*, Rivers Oram Press/ Pandora List.

7 See Moore, Wendy. (2016) Dr James Barry: A Woman Ahead of Her Time Review – an Exquisite Story of Scandalous Subterfuge, *Guardian*, 10 November. Accessed at: www.theguardian.com.

8 See: Holmes, Rachel. (2007) *The Secret Life of Dr James Barry: Victorian England's Most Eminent Surgeon*, London: The History Press Ltd; Duncker, Patricia. (2011) *James Miranda Barry*, London: Bloomsbury Paperbacks; and du Preez, Michael and Dronfield, Jeremy. (2016) *Dr James Barry: A Woman Ahead of Her Time*, Oneworld Publications

9 For instance, in 1896 *The Lady Cyclist* reported on how the popularity of the pneumatic tyre had 'opened up a new field of labour for women' as their 'fingers seem better fitted to joining and fitting' – see *The Lady Cyclist.* (1896) Tyre Making, 27 June, p. 164.

10 Eichhorn, Kate. (2013) *The Archival Turn in Feminism: Outrage in Order*, Pennsylvania: Temple University Press, p. 9.

11 Hemmings, Clare. (2011) *Why Stories Matter: The Political Grammar of Feminist Theory*, Duke University Press.

12 Bijker, Wiebe. (1995) *Of Bicycles, Bakelites and Bulbs: Toward a Theory of Sociotechnical Change*, Cambridge, MA: MIT Press, p. 1.

13 Star, Susan Leigh. (1999) The Ethnography of Infrastructure, *American Behavioral Scientist*, 43 (3): 377–391 (p. 377).

14 While writing this book, two women have appeared – Jane Austen is on the new ten-pound note and a statue of Millicent Fawcett will be the first of a woman in London's Parliament Square.

15 Willard, Frances. (1895) *A Wheel Within a Wheel: How I learned to ride the bicycle, with some reflections along the way*, New York: Flemming H. Revell Company.

16 'This Girl Can' campaign was launched in 2015, and supported by The National Lottery and Sport England. It is an example of how present these tropes are and how they are (still) being challenged. 'This Girl Can is a celebration of active women who are doing their thing no matter how well they do it, how they look or even how red their face gets'. See www.thisgirlcan. co.uk/.

17 While Olympic track events have reached gender parity – achieved in 2012 – other forms of cycle racing have not. The immensely popular Tour de France is for men only. Women are present on stage as presentation hostesses, more commonly known as podium girls, who pose in press photos with male winners. At the time of writing, the biggest international road race for elite women, the Giro Rosa, which has been running since 1988, was still not being covered by an English-language broadcaster. Cycling's worldwide governing body, the UCI (Union Cycliste Internationale), currently only legislates minimum wage for professional male cyclists. And amid post-London Olympic celebrations, British cycling has been rocked by allegations of sexism and discrimination towards elite women athletes.

18 See: 101Wankers (a map of sweary male car drivers – this map is no longer online but an article by Dawn Foster explains the project: www.theguardian. com/environment/green-living-blog/2010/aug/18/cycling-sexist-abuse-female), Hollaback (a non-profit movement about public harassment – www.ihollaback.org/) and the Near Miss Project (a study into non-injurious cycling incidents – www.nearmiss.bike/) are just a few projects that have documented anti-social behaviour towards mobile women in public.

Bibliography

Amram, Fred. (1984) The Innovative Woman, *New Scientist*, 24 May, p. 10.

Ashelford, Jane. (1996) *The Art of Dress: Clothes and Society*, London: The National Trust.

Atkins, Peter. (1990) The Spatial Configuration of Class Solidarity in London's West End 1792–1939, *Urban History Yearbook*, 17: 35–65.

The Australasian. (1896) Social Notes, 12 December, p. 42. Accessed at the National Library of Australia: http://trove.nla.gov.au.

The Australasian. (1897) Indispensible to Cyclists, Jaeger Advertisement, 16 January, p. 141.

Bassuk, Ellen L. (1985) The Rest Cure: Repetition or Resolution of Victorian Women's Conflicts? In: Rubin Suleiman, Susan. (ed.) *The Female Body in Western Culture: Contemporary Perspectives*, Cambridge, MA and London: Harvard University Press, pp. 139–151.

Beaudry, Mary Carolyn. (2006) *Findings: The Material Culture of Needlework and Sewing*, Yale University Press.

Beaujot, Ariel. (2002) *Victorian Fashion Accessories*, London and New York: Berg.

Bicycling News. (1893) Our Contemporaries, 7 October, p. 228. Accessed at the British Library.

Bicycling News and Sport and Play. (1985) Lines for Ladies by Marguerite, 16 April, p. 18.

Bicycling News and Sport and Play. (1985) Lines for Ladies by Marguerite, 7 May, p. 36.

Bicycling News and Sport and Play. (1895) Lines for Ladies by Marguerite, 14 May, p. 33.

Bicycling News and Sport and Play. (1895) Lines for Ladies by Marguerite, 28 May, p. 34, quoting from 'The Compromises of Cycling' in *Hearth and Home*.

Bicycling News and Sport and Play. (1985) Lines for Ladies by Marguerite, 16 April, p. 18.

Bicycling News and Sport and Play. (1895) Lines for Ladies by Marguerite, 4 June, p. 33.

Bicycling News and Sport and Play. (1895) Lines for Ladies by Marguerite, 18 June, p. 21.

Bicycling News and Sport and Play. (1895) 24 September, p. 10.

Bicycling News and Sport and Play. (1895) Patents, 19 November, p. 48.

Bicycling News and Tricycling Gazette. (1893) Ladies Page, 9 September, p. 168. Accessed at the British Library

Bicycling News and Tricycling Gazette. (1893) Women Awheel, 23 September, p. 1.

Bicycling News and Tricycling Gazette. (1893) The Ladies Page, 23 September, p. 198.

Bicycling News and Tricycling Gazette. (1893) Rational Dress for Ladies by Lacy Hiller, G, 30 September, p. 213.

Bicycling News and Tricycling Gazette. (1893) The Costume of the Future, The Ladies Page, 30 September, p. 215.

Bicycling News and Tricycling Gazette. (1893) The Ladies Page, 18 November, p. 2.

Bicycling News and Tricycling Gazette. (1893) Dress Reform For Women, by Ada Earland, 18 November, p. 6.

Bicycling News and Tricycling Gazette. (1893) Dress Reform in Peril, 23 December, p. 1.

Bijker, Wiebe. (1995) *Of Bicycles, Bakelites and Bulbs: Toward a Theory of Sociotechnical Change*, Massachusetts: MIT Press.

Bowker, Geoff and Star, Susan. (2000) *Sorting Things Out: Classification and its Consequences*, Cambridge, MA: MIT Press.

Breward, Christoper, Conekin, Becky and Cox, Caroline. (eds) (2002) *The Englishness of English Dress*, Berg.

The Buckman Papers. (1894–1900) Part of the Papers of Sydney and Maude Buckman relating to the Rational Dress Movement and Cycling for Women, accessed at the Hull History Centre. Ref: U DX113.

Bullock Workman, Fanny. (1896) Notes of a Tour in Spain, *The Lady Cyclist*, March, pp. 42–44. Accessed at the National Cycling Archive, Warwick University.

Burman, Barbara. (ed.) (1999) *The Culture of Sewing: Gender, Consumption and Home Dressmaking*, Oxford: Berg.

Burman, Barbara. (2002) Pocketing the Difference: Gender and Pockets in Nineteenth-Century Britain, *Gender and History*, 14 (3): 447–469.

Burman, Barbara and Denbo, Seth. The History of Pockets. Accessed at: www.vads.ac.uk/texts/POCKETS/history_of_tie-on_pockets.pdf.

Burman, Barbara, and Denbo, Seth. (2007) *Pockets of History: The Secret Life of an Everyday Object*, Museum of Costume Bath.

Burman, Barbara and Turbin, Carole. (2002) Introduction: Material Strategies Engendered, *Gender & History*, 14 (3): 371–381.

Burman, Barbara and Turbin, Carole. (2003) *Material Strategies: Dress and Gender in Historical Perspective*, Oxford: Blackwell.

Cambridge Orlando Project: Women's writing in the British Isles from the Beginning to the Present. Henrietta Muller, http://orlando.cambridge.org.

Campbell Warner, Patricia. (2006) *When the Girls Came Out to Play: The Birth of American Sportswear*, Amherst and Boston: University of Massachusetts Press.

The Church Weekly. (1897) Inventions Which Have Made People Rich, 5 February, p. 11. Accessed at: www.newspapers.com.

The Church Weekly. (1899) What Women Have Done, 6 October, p. 20.

Clarsen, Georgina. (2011) *Eat My Dust: Early Women Motorists*, JHU Press.

The Comptroller General of Patents, Designs, and Trade Marks, with Appendices Annual Reports, 1884–1898.

Constanzo, Marilyn. (2010) 'One can't shake off the women': Images of Sport and Gender in *Punch*, 1901–10, *The International Journal of the History of Sport*, 19 (1): 31–56 (p. 35).

Crane, Diana. (2000) *Fashion and its Social Agendas: Class, Gender, and Identity in Clothing*, Chicago and London: University of Chicago Press.

Crawford, Elizabeth. (2003) *The Women's Suffrage Movement: A Reference Guide 1886–1928*, London: Routledge.

Crawford Munro, J. E. (1884) *The Patents, Designs, and Trade Marks Act, 1883 (46 & 47 Vict. C. 57) with the Rules and Instructions, Together with pleadings, order and precedents*, London, p. xxxvii. Accessed at: https://archive.org/stream/patentsdesignsa01britgoog#page/n6/mode/2up.

Cresswell, Tim. (1999) Embodiments, Power and the Politics of Mobility: The Case of Female Tramps and Hobos, *Transactions of the Institute of British Geographers*, 24 (2): 175–192.

CTC Monthly Gazette. (1882) Excelsior Tricycle, Advertisement, February. Accessed at Cycling UK.

CTC Monthly Gazette. (1883) The Ladies Dress, June, p. 270.

CTC Monthly Gazette. (1883) The New Uniform, July, p.292.

CTC Monthly Gazette. (1883) The Ladies Dress, July, p. 293.

CTC Monthly Gazette. (1883) The Ladies Dress, August, p. 320.

CTC Monthly Gazette. (1883) The New Uniform, August, p. 313.

CTC Monthly Gazette. (1883) The Ladies Dress. Important, December, p. 392.

CTC Monthly Gazette. (1886) A Growing Abuse, August, p. 322.

CTC Monthly Gazette. (1894) The Ladies' Page, February, p. 62.

CTC Monthly Gazette. (1897) Box O' Lights My Lord, January, p. 9.

CTC Monthly Gazette. (1898) March, p. 10.

The Cycling World Illustrated. (1896) A Sunny Morning in Battersea Park, 18 March, p. 8.

The Cycling World Illustrated. (1896) Cycling Whispers, 25 March, p. 32.

The Cycling World Illustrated. (1896) Distinguished Lady Cyclists – Mrs Houston French, 25 March, p. 30.

The Cycling World Illustrated. (1896) Miss Vigor, Photo by Faulkner, Baker Street, W., 27 May, p. 241.

The Cycling World Illustrated. (1896) 3 June, p. 269.

The Cycling World Illustrated. (1896) A Parisian Divided Skirt, 3 June, p. 273.

The Cycling World Illustrated. (1896) Array Yourself Becomingly, 24 June, p. 351.

The Cycling World Illustrated. (1896) Chats With Cyclists – A Cyclist's Chaperone: Miss Cave by Mr Bancroft, 24 June, p. 342.

Daily Telegraph. (1896) Cyclists' Association – Skirts V. Rationals, 28 November. In The Buckman Papers (1894–1900).

The Dawn: A Journal for Australian Women. (1896) Answers to Correspondents, 1 November. Accessed at The National Library of Australia: http://trove.nla. gov.au.

de Laet, Marianne and Mol, Annemarie. (2000) The Zimbabwe Bush Pump: Mechanics of a Fluid Technology, *Social Studies of Science*, 30: 225-263.

Dolan, Brian. (2001) *Ladies of the Great Tour*, London: HarperCollins.

Duncker, Patricia. (2011) *James Miranda Barry*, London: Bloomsbury Paperbacks.

du Preez, Michael and Dronfield, Jeremy. (2016) *Dr James Barry: A Woman Ahead of Her Time*, Oneworld Publications.

Eichhorn, Kate. (2013) *The Archival Turn in Feminism: Outrage in Order*, Pennsylvania: Temple University Press.

Erskine, F. J. (2014) *Lady Cycling: What to Wear & How to Ride*, first published in 1897 by Walter Scott Ltd, London: The British Library.

Evans, Hilary and Evans, Mary. (1976) *The Party that Lasted 100 Days: The Late Victorian Season: A Social Study*, London: Macdonald and Jane's, p. 131.

Evening Star (Washington, District of Columbia). (1904) Miss F. Henrietta Muller. B.A will give three informal talks to men and women in the Assembly Hall at the Shoreham Hotel, 12 May, p. 12. Accessed at: www.newspapers.com.

Evening Star (Washington, District of Columbia). (1904) Amusements – Two Lectures will be delivered by Miss. F. Henrietta Muller on Free Motherhood or Parthenogenesis, 12 February, p. 16. Accessed at: www.newspapers.com.

Fennetaux, Ariane. (2008) Women's Pockets and the Construction of Privacy in the Long Eighteenth Century, *Eighteenth-Century Fiction*, 20 (3): 307-334.

Gamber, Wendy. (1997) *The Female Economy: The Millinery and Dressmaking Trades, 1860-1930*, Urbana: University of Illinois Press.

Godley, Andrew. (2006) Selling the Sewing Machine Around the World: Singer's International Marketing Strategies, 1850-1920, *Enterprise & Society*, 7 (2): 266-314.

Gordon, Sarah. (2001) 'Any Desired Length': Negotiating Gender through Sports Clothing, 1870-1925. In Scranton, Philip. (ed.) *Beauty and Business: Commerce, Gender and Culture in Modern America*, New York and London: Routledge, pp. 24-51.

Gordon, Sarah. (2007) *'Make it Yourself': Home Sewing, Gender and Cultures 1890-1930*, Columbia University Press. Accessed at: www.gutenberg-e.org/gordon/.

Greville, Violet. (1894) *Ladies in the Field: Sketches of Sport*, London: Ward & Downy, p. 261.

Guardian. (2010) Divorce Rates Data, 1858 to Now: How Has it Changed? *Guardian News Datablog*, 28 January. Accessed at: www.theguardian.com.

Hammersley, Martyn and Atkinson, Paul. (2007) *Ethnography: Principles in Practice*, Third Edition, Oxon: Routledge.

Hanlon, Sheila. (2011) Lady Cyclist Effigy at the Cambridge University Protest, 1897. Research website, posted 16 April. Accessed at: www.sheilahanlon.com.

Hanlon, Sheila. (2011) The Way to Wareham: Lady Cyclists in Punch Magazine Cartoons, 1890s. Research website, posted 7 April. Accessed at www.sheilahanlon.com/.

Hanlon, Sheila. (2013) Cycling to Suffrage: Bicycles and the Organised Women's Suffrage Movement in Britain, 1900–1914, *Cycle History*.

Hargreaves, Jennifer. (1994) *Sporting Females: Critical Issues in the History and Sociology of Women's Sports*, London and New York: Routledge.

Hargreaves, Jennifer. (2001) The Victorian Cult of the Family and the Early Years of Female Sport. In Scraton, Sheila and Flintoff, Anne. (eds) *Gender and Sport: A Reader*, London and New York: Routledge, pp. 53–65.

Helvenston Gray, Sally and Peteu, Michaela C. (2005) 'Invention, the Angel of the Nineteenth Century': Patents for Women's Cycling Attire in the Nineteenth Century, Dress, 32: 27–42.

Hemmings, Clare. (2011) *Why Stories Matter: The Political Grammar of Feminist Theory*, Durham, NC: Duke University Press.

Holcombe, Lee. (1973) *Victorian Ladies at Work: Middle-Class Working Women in England and Wales 1850–1914*, Connecticut: Archon Books.

Hollaback! A non-profit movement to end harassment in public spaces. Accessed at: www.ihollaback.org/.

Holmes, Rachel. (2007) *The Secret Life of Dr James Barry: Victorian England's Most Eminent Surgeon*, The History Press Ltd.

Horton, D., Rosen, P. and Cox, P. (eds) (2016) *Cycling and Society*, London and New York: Routledge (first published by Ashgate in 2007).

Huang, Jinya. (2007) Queen of the Road: Bicycling, Femininity, and the Lady Cyclist, *Cycle History*, 17: 69–76.

The Hub. (1896) Pars from the Parks, 15 August, p. 67. Accessed at the National Cycling Archive, Warwick University.

The Hub. (1896) Women More Careful, 22 August, p. 104.

The Hub. (1896) The Tables are Now Turned, 19 September, p. 255.

The Hub. (1896) Danger in Excess: Most Women Try to Ride Too Long and Too Hard, 19 September, p. 207.

The Hub. (1896) Should Women Cycle: Medical View of the Question, 19 September, p. 269.

Jerome, Jerome K. (2016) *Women and Wheels (written in 1897).* In Spencer, Charles., Pain, Barry., Jerome, Jerome, K., and Herschell, George. *Cycling: The Craze of the Hour*, Pushkin Press, pp. 82–91.

Jungnickel, Kat. (2015) 'One needs to be very brave to stand all that': Cycling, Rational Dress and the Struggle for Citizenship in Late Nineteenth Century Britain, *Geoforum, Special Issue: Geographies of Citizenship and Everyday (Im) Mobility*, 64: 362–371.

Jungnickel, Kat. (2017) Making Things To Make Sense of Things: DiY as Research and Practice. In Sayer, J. (ed.) *The Routledge Companion to Media Studies & Digital Humanities*, London: Routledge.

Kinsey, Fiona. (2012) Reading Photographic Portraits of Australian Women Cyclists in the 1890s: From Costume and Cycle Choices to Constructions of Feminine Identity. In Huggins, Mike and O'Mahony, Mike. (eds) *The Visual in Sport*, London and New York: Routledge, pp. 35–51.

The Ladies Tailor. (1896) Plate 2. January, p. 7. Accessed at The British Library.

The Ladies' Tailor. (1897) The Ladies' Tailor Spring Fashions 1897, January, No. 1, Vol. 13.

The Ladies' Tailor. (1897) Illustrations of British Costumes, Spring 1897, March, No. 3, Vol. 13, p. 25.

The Ladies' Tailor. (1897) Notes of the Month, Spring 1897, March, No. 3, Vol. 13, p. 26.

The Ladies' Tailor. (1898) Illustrations of Ladies' Cycling Costume, March, No. 3, Vol. 14, p. 26.

The Lady Cyclist. (1896) Advertisement for the Fixit Dressholder, March, p. 18. Accessed at Manchester Art Gallery archives (Gallery of Costume, Platt Hall).

The Lady Cyclist. (1896) Letters of a Lady Cyclist, March, p. 20.

The Lady Cyclist. (1896) Lady Cyclists at Home – Mrs Selwyn F. Edge, March, p. 27.

The Lady Cyclist. (1896) Why Lady Cyclists Should Always Dress Well, March, p. 30.

The Lady Cyclist. (1896) March, p. 38.

The Lady Cyclist. (1896) Repaired Her Punctured Dunlop Tyre in 5 Minutes, March, p. 43.

The Lady Cyclist. (1896) A Novel Skirt, March, p. 46.

The Lady Cyclist. (1896) 'En Avant' Cycling Skirt Advertisement, March, p. 46.

The Lady Cyclist. (1896) All Wool, March, p. 49.

The Lady Cyclist. (1896) The Thomas Skirt, March, p. 56.

The Lady Cyclist. (1896) Advertisement: John Gooch – Ladies Tailor and Cycle Habit Maker, March, back page.

The Lady Cyclist. (1896) Cycling Fashions by Madame Mode, 27 June, p. 167.

The Lady Cyclist. (1896) Tyre Making, 27 June, p. 164.

The Lady Cyclist. (1896) The Paris Mode – The Jupe-Culotte, 27 June, p. 172.

The Lady Cyclist. (1896) No Wonder! 22 August, p. 419.

The Lady Cyclist. (1896) Cycling Fashions by Madame Mode, 22 August, p. 426.

The Lady Cyclist. (1896) Lady Cyclists At Home – Mrs Paul Hardy, 22 August, pp. 433–434.

The Lady Cyclist. (1896) Famous Cycle Emporiums, 22 August, p. 438.

The Lady Cyclist. (1896) The New Woman, 22 August, p. 430.

The Lady Cyclist. (1896) A Mistaken Idea, 29 August, p. 451.

The Lady Cyclist. (1896) Lady Snap-shotters, 26 September, p. 580.

The Lady Cyclist. (1896) Dress Well, 26 September, p. 581.

The Lady Cyclist. (1896) *Little Essay's by Cynthia, On Dress-Guards*, 24 October, p. 719.

The Lady Cyclist. (1896) Editorial, 31 October.

The Lady Cyclist. (1896) PHOTOGRAPHS! The Prettiest Cyclist in the Prettiest Costume! A novel form of prize, 31 October, p. 761.

The Lady's Newspaper. (1895) Wayside Jottings, by A. Wheeler (pseudonym of S. S. Buckman), July, p. 177. Accessed at Manchester Art Gallery archives (Gallery of Costume, Platt Hall).

The Lady's Own Magazine. (1897) Wayside Jottings, October, p. 119.

The Lady's Own Magazine. (1898) Wayside Jottings, March, p. 83.

The Lady's Own Magazine. (1898) A Clipping from an American Paper, June, p. 173.

The Lady's Own Magazine. (1898) The Modern Matchmaker, June, p. 173.

The Lady's Own Magazine. (1898) Readily Understood, August, p. 51.

Lampland, Martha and Star, Susan Leigh. (eds) (2009) *Standards and Their Stories: How Quantifying, Classifying and Formalising Practices Shape Everyday Life*, Ithaca and London: Cornell University Press.

Latour, Bruno and Woolgar, Steve. (1979) *Laboratory Life: The Social Construction of Scientific Facts*, London: Sage.

Lewis, Aimee. (2014) Is Sport Sexist? Six Sports where Men and Women are Still Set Apart. BBC Sport, 18 September. Accessed at: www.bbc.co.uk/sport/golf/29242699.

The London Illustrated. (1895) London Season, 1 June, p. 15. Accessed at the British Library.

Longhurst, Robyn. (2001) *Bodies: Exploring Fluid Boundaries*, London: Routledge.

The Manufacturer and Inventor, London. (1891) Wages in the Minor Textile Trades, 20 February, p. 64, Accessed at: www.britishnewspaperarchive.co.uk.

Maudsley, Henry. (1894) Sex in the Mind and in Education, *Popular Science Monthly*, 5, June.

McCrone, Kathleen. (1988) *Sport and the Physical Emancipation of English Women, 1870–1914*, London: Routledge.

McLachlan, Fiona. (2016) Gender Politics, the Olympic Games and Road Cycling: A Case for Critical History, *The International Journal of the History of Sport*, 33 (4): 469–483.

McWilliams-Tullberg, Rita. (1998) *Women at Cambridge*, Cambridge: Cambridge University Press.

Miller, Daniel, Jackson, Peter, Thrift, Nigel, Holbrook, Beverley and Rowlands, Michael. (1998) *Shopping, Place and Identity*, Oxon: Routledge.

Moore, Wendy. (2016) Dr James Barry: A Woman Ahead of Her Time Review – an Exquisite Story of Scandalous Subterfuge, *Guardian*, 10 November. Accessed at: www.theguardian.com.

The Morning Leader. (1899) A Rational Cycling Costume – A Costume for the Country, 6 February. In The Buckman Papers (1894–1900).

New York Sunday World. (1896) Champion of Her Sex: Miss Susan B. Anthony Tells the Story of Her Remarkable Life to Nellie Bly, 2 February, p. 10. Accessed at: www.nytimes.

New York Times. (1894) The Wheelmen of the Day; How Rapidly They Are Multiplying and a Word of Their Dress, 12 August, p. 18.

New York Times. (1897) Women and Cycling by Ida Trafford-Bell, *New York Times* Supplement, Sunday 7 February, p. 11. Accessed at: www.nytimes.

New York Times. (1897) Woman and the Bicycle: Mrs Hudders's Dress and Wheel, 11 October, p. 4.

Office for National Statistics; National Records of Scotland; Northern Ireland Statistics and Research Agency. (1901) Census aggregate data. UK Data Service. Accessed at: www.ons.gov.uk.

Office for National Statistics; National Records of Scotland; Northern Ireland Statistics and Research Agency. (1911) Census aggregate data. UK Data Service. Accessed at: www.ons.gov.uk.

Oddy, Nicholas. (2010) Rides on My Safety. The Diary of Emily Sophia Coddington (National Cycle Archive MSS.328/N28). In *Cycle History*, Birmingham: JPMPF, pp. 29–34.

Parker, Claire. (2010) Swimming: The Ideal 'Sport' for Nineteenth-Century British Women, *The International Journal of the History of Sport*, 27 (4): 675–689.

Parker, Rosika. (2012) *The Subversive Stitch: Embroidery and the Making of the Feminine*, London and New York: I. B. Tauris.

Peiss, Kathy. (2001) On Beauty … and the History of Business. In Scranton, Philip. (ed.) *Beauty and Business: Commerce, Gender and Culture in Modern Americ*a, New York and London: Routledge, pp. 7–23.

Pennell, Elizabeth R. (1897) Around London by Bicycle, *Harper's Monthly Magazine*, 95 (June): 489–510.

Perkins Gilman, Charlotte. (1997) If I were a Man. In *The Yellow Wallpaper and Other Stories*, Dover Thrift Editions, p. 58.

Peteu, Michaela Cornelia. (2004). Fashion and Function in Women's Dress as Revealed in Clothing Patents, 1846-1920. Unpublished doctoral dissertation, Michigan State, University, East Lansing.

Peteu, Michaela Cornelia and Helveston Gray, Sally. (2009) Clothing Invention: Improving the Functionality of Women's Skirts, 1846–1920, *Clothing & Textiles Research Journal*, 27 (1): 45–61.

Post Office London Directory. (1895) *Official Street, Commercial, Trade, Law, Court, Parliamentary, Postal. City and Clerical, Conveyance and Banking Directories*, 96th Annual Publication, London: Kelly & Son Ltd.

Punch. (1895) The Fate of Rotten Row, 13 July, p. 23. Accessed at Manchester Art Gallery archives (Gallery of Costume, Platt Hall).

Punch. (1896) Rational Costume – The Vicar of St. Winifred-in-the-Wold, 13 June, p. 282.

The Queen, the Lady's Newspaper. (1895) Dress Echoes of the Week, 30 March, p. 560. Accessed at Manchester Art Gallery archives (Gallery of Costume, Platt Hall).

The Queen, the Lady's Newspaper. (1895) The Weight of Ladies' Bicycles, 13 July, p. 85.

The Queen, the Lady's Newspaper. (1895) Dress Echoes of the Week, 16 November, p. 923.

The Queen, the Lady's Newspaper. (1895) Ladies' Outdoor Amusements, 7 December, p. 1095.

The Queen, the Lady's Newspaper. (1896) No. 1. French Visiting Dress, 4 January, p. 25.

The Queen, the Lady's Newspaper. (1896) No. 4. Visiting Gown, 11 January, p. 73.

The Queen, the Lady's Newspaper. (1896) Hyde Park on Sunday, 18 January, p. 89.

The Queen, the Lady's Newspaper. (1896) Women as Patentees, 18 January, p. 104.

The Queen, the Lady's Newspaper. (1896) The Culture of the Cycle, 25 January, p. 169.

The Queen, the Lady's Newspaper. (1896) Bicycling in Hyde Park, 8 February, p. 258.

The Queen, the Lady's Newspaper. (1896) Cycling News, 21 March, p. 515.

The Queen, the Lady's Newspaper. (1896) The Bygrave 'Quick Change' Cycling Skirt, 4 April, p. 595.

The Queen, the Lady's Newspaper. (1896) Mr Marcus' New Cycling Room, 4 April, p. 598.

The Queen, the Lady's Newspaper. (1896) Fashionable Capes, Pardessus and Millinery at Messrs Redmayne's Bond Street, 2 May, p. 756.

The Queen, the Lady's Newspaper. (1896) Cycling, 16 May, p. 886.

The Queen, the Lady's Newspaper. (1896) The New Cycling Costumes of the Season, by Various Makers, 11 July, p. 83.

The Queen, the Lady's Newspaper. (1897) George Lichtenfeld – The 'Pneumatic Tube Coil' – Latest Novelty, Advertisement, 19 June. Accessed at Manchester Art Gallery archives (Gallery of Costume, Platt Hall).

Rappaport, Erika Diane. (2000) *Shopping for Pleasure: Women in the Making of London's West End*, New Jersey and West Sussex: Princeton University Press.

The Rational Dress Gazette, Organ of the Rational Dress League. (1898) Editorial, June, No. 1, p. 2. Accesssed at The University of Hull.

The Rational Dress Gazette, Organ of the Rational Dress League. (1898) The Rational Dress League, June, No. 1, p. 4.

The Rational Dress Gazette, Organ of the Rational Dress League. (1899) A Weather Forecast, Illustration, No. 6, February, p. 17.

The Rational Dress Gazette, Organ of the Rational Dress League. (1899) Notes and Comments, No. 7, April, p. 25.

The Rational Dress Gazette, Organ of the Rational Dress League. (1899) Supplement to the Rational Dress Gazette, April, No. 7, pp. 1–4.

The Rational Dress Gazette, Organ of the Rational Dress League. (1899) No. 8, May, p. 29.

The Rational Dress Gazette: Organ of the Rational Dress League. (1899) Correspondence from Madel Sharman Crawford, June, No. 9, p. 36.

The Rational Dress Gazette, Organ of the Rational Dress League. (1899) A Skirt Responsible for 'Serious Injuries', No. 9, June, p. 86.

The Rational Dress Gazette, Organ of the Rational Dress League. (1899) Rational Underclothing, No. 10, July, p. 38.

The Rational Dress Gazette, Organ of the Rational Dress League. (1899) *Correspondence – Irene Marshall*, No. 10, July, p. 40.

The Rational Dress Gazette, Organ of the Rational Dress League. (1899) The Womanly and Graceful Skirt, No. 11, August, p. 19.

The Rational Dress Gazette, Organ of the Rational Dress League. (1899) The Fatal Skirt, No. 12, September, p. 46.

The Rational Dress Gazette, Organ of the Rational Dress League. (1899) Tailor Made, No. 13, October, p. 52.

The Rational Dress Gazette, Organ of the Rational Dress League. (1899) Editorial, No. 14, November, p. 54.

Robins Pennell, Elizabeth. (1897) Around London by Bicycle, *Harper's Monthly Magazine*, 95 (June): 489–510.

Robinson, J. (1747) *The Art of Governing a Wife; with rules for bachelors. To which is added, an essay against unequal marriages*, London: Gale ECCO.

Robinson, Jane. (2009) *Bluestockings: The Remarkable Story of the First Women to Fight for an Education*, London: Penguin. Russett, Cynthia E. (1989) *Sexual*

Science: The Victorian Construction of Womanhood, Cambridge, MA: Harvard University Press.

The Saint Paul Daily Globe. (1896) $5000 in Four Days: Mrs Bygrave received it last week for a bicycle skirt, 15 March, p. 19. Accessed at the British Newspaper Archive: www.britishnewspaperarchive.co.uk.

San Francisco Chronicle. (1896) For the Girl Who Would Go-A-Wheeling, 12 April, p. 8. Accessed at: www.britishnewspaperarchive.co.uk.

Schwartz-Cowan, Ruth. (1979) From Virginia Dare to Virginia Slims: Women and Technology in American Life, *Technology & Culture*, 20 (1): 51–63.

Schwartz-Cowan, Ruth. (1983) *More Work for Mother: The Ironies of Household Technology from the Open Hearth to the Microwave*, New York: Basic Books.

Schwartz-Cowan, Ruth. (1997) Inventors, Entrepreneurs and Engineers. In *A Social History of American Technology*, New York and Oxford: Oxford University Press, pp. 119–148. *Scientific American*. (1870) Female Inventive Talent, 17 September, p. 184.

Showalter, Elaine. (1987) *The Female Malady: Women, Madness and English Culture, 1830–1980*, London: Virago.

Simpson, Clare. (2001) Respectable Identities: New Zealand Nineteenth-Century 'New Women' – on Bicycles!, *The International Journal of the History of Sport*, 18 (2): 54–77.

Simpson, Clare. (2016) Capitalising on Curiosity: Women's Professional Cycle Racing in the Late-Nineteenth Century. In Horton, D., Rosen, P. and Cox, P. (eds) *Cycling and Society*, London and New York: Routledge (first published by Ashgate in 2007), pp. 47–66.

Skeggs, Bev. (2007) Feminist Ethnography. In Atkinson, P., Coffey, A. J., Delamont, S., Lofland, J. and Lofland, L. H. (eds) *Handbook of Ethnography*, London: Sage, pp. 426–442.

The Sketch. (1896) Society on Cycles, 22 January, p. 686. Accessed at Manchester Art Gallery archives (Gallery of Costume, Platt Hall).

The Sketch. (1896) Women Cyclists at the Agricultural Hall, 8 April, p. 453.

The Sketch. (1896) In Hyde Park, The Cycling Craze, 22 August, p. 543.

The Sketch. (1896) The Lady Cyclists at the Aquarium, 27 November, p. 233.

The Sketch. (1896) Society on Wheels, 11 March, p. 311.

Smethurst, Paul. (2015) *The Bicycle: Towards a Global History*, London: Palgrave Macmillan.

St Louis Post Dispatch. (1896) Mrs Bygrave's Bicycle Skirt, 8 March, p. 22. Accessed at: www.britishnewspaperarchive.co.uk.

Stanley, Autumn. (1998) *Mothers and Daughters of Invention: Notes for a Revised History of Technology*, New Brunswick: Rutgers University Press.

Stanley, Jo. (1995) *Bold in Her Breeches: Women Pirates Across the Ages*, Hammersmith: HarperCollins.

Stoler, Ann Laura. (2002) Colonial Archives and the Arts of Governance, *Archival Science*, 2: 87–109.

Star, Susan Leigh. (1999) The Ethnography of Infrastructure, *American Behavioral Scientist*, 43 (3): 377–391.

Swanson, Kara W. (2009) The Emergence of the Patent Practitioner, *Technology & Culture*, 50 (3): 519–548.

Sweet, Matthew. (2001) *Inventing the Victorians*, London: Faber & Faber.

The Sydney Daily Mail and New South Wales Advertiser. (1896) Advertisement for the Bygrave Convertible Skirt, 7 November, p. 992. Accessed at the National Library of Australia: http://trove.nla.gov.au.

The Sydney Daily Mail and New South Wales Advertiser. (1896) *The Ladies' Page*, 5 December, p. 1200.

The Sydney Morning Herald. (1896) 26 September, p. 5. Accessed at The National Library of Australia: http://trove.nla.gov.au.

The Tailor and Cutter. (1895) Fashions in the Park, 2 May, p. 174. Accessed at Manchester Art Gallery archives (Gallery of Costume, Platt Hall).

The Tailor and Cutter. (1895) Cycling Garments at the Stanley Show, 5 December, p. 485.

The Tailor and Cutter. (1896) Goods From Shop Windows, 9 January, p. 20.

The Tailor and Cutter. (1896) The Slack Season, Vol. 31, No. 1,532, 13 February, p. 67.

Tamboukou, Maria. (2016) *Sewing, Fighting and Writing: Radical Practices in Work, Politics and Culture*, London and New York: Rowman & Littlefield.

Tickner, Lisa. (1987) *The Spectacle of Women: Imagery of the Suffrage Campaign 1907–14*, London: Chatto & Windus.

The Times. (1881) Parliamentary Intelligence, House of Commons – Patents for Invention Bill, 15 June, p. 6.

The Times. (1899) Messrs. Staley, Popplewell and Co. Chartered Patent Agents from 61 Chancery Lane, 6 January, p. 15. Accessed at the British Newspaper Archive: www.britishnewspaperarchive.co.uk.

The Times. (1899) 9 March, p. 14.

Towzer, Jane and Levitt, Sarah. (2010) *Fabric of Society: A Century of People and their Clothes 1770–1870*, The Gallery of Costume, Manchester, printed in association with Laura Ashley Publications.

Veblen, Thorstein. (1934) *The Theory of the Leisure Class*, London.

Vertinsky, Patricia. (1990) *Eternally Wounded Woman: Women, Doctors, and Exercise in the Late Nineteenth Century*, Urbana and Chicago: University of Illinois Press.

Wajcman, Judy. (2004) *Technofeminism*, Cambridge: Polity Press.

Walkowitz, J. R. (1992) *City of Dreadful Delight: Narratives of Sexual Danger in Late-Victorian London*, Chicago: University of Chicago Press.

Weir Mitchell, Silas. (1877) *Fat and Blood and How to Make Them*, Philadelphia: J. B. Lipincott & Co.

Wells, H. G. (2016) *Time Machine*, Wisehouse Classics (originally published in 1895).

The Westminster Budget. (1896) Concerning Hearth and Home: Wear on the Wheel, 3 April, p. 28. Accessed at: www.newspapers.com.

The Wheeler. (1894) Mrs Barrington, No. 115, Vol. 5, 27 June, front page.

The Wheelwoman and Society Cycling News. (1896) Jaegar Advert, 28 November, p. 34. Accessed at The British Library.

The Wheelwoman and Society Cycling News. (1897) A New Saddle Cover, 23 January, p. 6.

The Wheelwoman and Society Cycling News. (1897) When in Doubt – Consult Cynthia, 11 September, p. 19.

Wheelwright, Julie. (1989) *Amazons and Military Maids: Women Who Dressed as Men in Pursuit of Life, Liberty and Happiness*, Rivers Oram Press/ Pandora List.

Willard, Frances. (1895) *A Wheel Within a Wheel: How I Learned to Ride the Bicycle, With Some Reflections Along the Way*, New York: Flemming.H. Revell Company.

Willett Cunnington, Cecil. (2013) *English Women's Clothing in the Nineteenth Century: A Comprehensive Guide*, Dover Publications.

Woman's Herald: Women's Penny Paper. (1888) No. 1, Vol. 1, 27 October, p. 4.

Woman's Herald: Women's Penny Paper. (1888) No. 2, Vol. 1, 3 November, p. 2.

Woman's Herald: Women's Penny Paper. (1888) No. 6, Vol. 1, 1 December, p. 2.

Woman's Herald: Women's Penny Paper. (1889) 2 March, p. 7.

Woman's Herald: Women's Penny Paper. (1891) Interview: Miss F. Henrietta Muller, Editor of the Woman's Herald, by Clara E. De Moleys, 28 November, p. 916. Accessed at the Women's Library, London School of Economics.

The Worker (Brisbane, Qld). (1896) Sporting. 23 May, p. 10. Accessed at the National Library of Australia: http://trove.nla.gov.au.

Zorina Khan, B. (2005) *The Democratization of Invention: Patents and Copyrights in American Economic Development, 1790–1920,* Cambridge: Cambridge University Press, p. 128.

Zorina Khan, B. (2000) 'Not for Ornament': Patenting Activity by Nineteenth Century Women Inventors, *Journal of Interdisciplinary History,* 31 (2): 159–195.

Figures

Index